# The Consumer's Guide to Wireless Security

7333
SEP 0 4 1998
INFORMATION RESOURCES CENTER
ASIS
1625 PRINCE STREET
ALEXANDRIA, VA 22314
tel. (703) 519-6200
TH
9745
.D8M67
1996

# The Consumer's Guide to Wireless Security

*Joseph Moses*
*Lou Sepulveda*

**McGraw-Hill**

New York San Francisco Washington, D.C. Auckland Bogotá
Caracas Lisbon London Madrid Mexico City Milan
Montreal New Delhi San Juan Singapore
Sydney Tokyo Toronto

**Library of Congress Cataloging-in-Publication Data**

Moses, Joseph (Joseph K.)
The consumer's guide to wireless security / Joseph Moses, Lou Sepulveda.
p. cm.
Includes index.
ISBN 0-07-043492-1 (hc.) — ISBN 0-07-043493-X (pbk.)
1. Dwellings—Security measures. 2. Electronic security systems.
I. Sepulveda, Lou. II. Title.
TH9745.D85M67 1996
643'.16—dc21 96-47048
CIP

*McGraw-Hill*

*A Division of The* ***McGraw-Hill*** *Companies*

Copyright © 1997 by The McGraw-Hill Companies, Inc. All rights reserved. Printed in the United States of America. Except as permitted under the United States Copyright Act of 1976, no part of this publication may be reproduced or distributed in any form or by any means, or stored in a data base or retrieval system, without the prior written permission of the publisher.

1 2 3 4 5 6 7 8 9 0 DOC/DOC 9 0 1 0 9 8 7 6

ISBN 0-07-043492-1 (HC)
0-07-043493-X (PBK)

The sponsoring editor for this book was Steven S. Chapman, the editing supervisor was Lori Flaherty, and the production supervisor was Suzanne W.B. Rapcavage. It was set in ITC Century Light by Jana Fisher through the services of Barry E. Brown (Broker—Editing, Design and Production).

Printed and bound by R.R. Donnelley & Sons Company.

McGraw-Hill books are available at special quantity discounts to use as premiums and sales promotions, or for use in corporate training programs. For more information, please write to the Director of Special Sales, McGraw-Hill, 11 West 19th Street, New York, NY 10011. Or contact your local bookstore.

Information contained in this work has been obtained by The McGraw-Hill Companies, Inc. ("McGraw-Hill") from sources believed to be reliable. However, neither McGraw-Hill nor its authors guarantees the accuracy or completeness of any information published herein and neither McGraw-Hill nor its authors shall be responsible for any errors, omissions, or damages arising out of use of this information. This work is published with the understanding that McGraw-Hill and its authors are supplying information but are not attempting to render engineering or other professional services. If such services are required, the assistance of an appropriate professional should be sought.

This book is printed on recycled, acid-free paper containing a minimum of 50% recycled, de-inked fiber.

# Contents

# Introduction

Wireless home security systems epitomize the magic of modern technology. They give homeowners the extraordinary power to take action in the face of devastating threats to life safety and security. Wireless security systems are the "Open Sesame" for the twenty-first century. They confer the power to protect homes against burglary or to summon help in the event of a fire, a break-in, a fall, or a heart attack. They're so user-friendly that even people with command-o-phobia—the irrational fear of PC pop-up menus—have the power to point-and-click without having to touch a mouse or even face a dreaded LCD. Perhaps most importantly, the wireless security system represents our personal power to fight crime. For a nation sick of violence, the wireless security system is the sword, the Winchester, the Colt 45, the light saber for the new millennium. Who needs brute force when you can activate a full-perimeter home security system from 500 feet away with a wireless keychain touchpad?

*The Consumer's Guide to Wireless Security* is a comprehensive resource for consumers searching for answers to questions about wireless security. You will learn what the components of a wireless security system are, how they work together to provide security protection, how to evaluate a system for specific needs, and how to choose a dealer/installer. The guide is designed for both the do-it-yourselfer and for the consumer interested in making an informed decision when choosing a security dealer.

The price of dealer-installed security systems has come down in the last few years, which is good news for everyone. It costs consumers less than ever to have detection security, and the market for security-system manufacturers has broadened. Where the cost of wireless systems was once prohibitive for all but a small elite, a wide range of systems with a variety of components is now within the means of most homeowners.

According to *Security Dealer* magazine, wireless technology is in its golden age. The technology has never been more reliable or easy to use. But this also means that more and more manufacturers and security dealers are springing up, hoping to cash in on the expanding security industry, and unfortunately, not all dealers and equipment manufacturers are created equal. This guide is designed to prepare you to make informed decisions about the best wireless electronic security components for your needs.

For decades, social commentators have been telling us that we live in an increasingly fast-paced society. The "wireless society" is faster than fast, and it's not on a pace at all—it's on a pulse, a wave, a signal that speeds along between 300 and 900 MHz. What does that mean? It means you no longer have to wait until after you walk into a darkened house to turn on the lights; you can have the place lit up in advance. With some systems, you can call up your wireless security system control panel from your workplace on a cold winter afternoon and turn up the heat from your Touch-Tone phone so the house is warmed up by the time you get there.

What makes wireless security technology better than hardwire? Think of the differences in communicative power and convenience between the telegraph and your cordless phone and you'll have a clear picture of the advantages of wireless technology. Wireless security is one of the most accessible high-tech tools available—and one of the most necessary. This technology allows you to control things from a distance—like locking and unlocking doors and turning lights off or on—in order to increase your safety and security.

Say, for instance, you are in your garden, and you see a burglary or assault in progress, you have a medical emergency, or you see a fire break out. With a wireless panic button, you can set off an alarm that will signal your control panel and notify the central station of an emergency.

With wireless technology, you have digital weapons for fighting crime that no criminal can take away and use against you; a touchpad is useless to anyone who doesn't know what buttons to push. You can arm and disarm your car's security system with an access code that no valet can take off a key-ring and have duplicated while you're at the theater. Besides, wires slow you down. It takes time to install wires for hardwire security systems—lots of time. Holes must be drilled, wire must be fished, conduit must be installed. And you'll need to take time off from work or other activities to watch over things while installers practically move into your house for a two- or three-day visit. Labor-intensive hardwire installations eat up labor hours and can create installation backlogs of days and weeks, so you might be put on a waiting list before work can even begin.

Wireless security systems, on the other hand, can be installed in four to eight hours, and components can be placed where they will be most efficient and where they are the most convenient for you to use—not just where it's easiest to run wires.

## About this book

This book will give you the background on the history of the fastest growing segment of the security industry and tell you exactly how wireless security works, what system components do and how they work, how security can be integrated with other home automation systems, how to choose a reliable security system and a dealer to install it, or how to plan, install, and test a wireless system yourself.

The appendix includes easy-to-use planning forms and worksheets for designing a system, a false-alarm prevention tip sheet, a fire-escape planning form, and other life safety ideas to protect your home and family. A comprehensive glossary is included so you can discuss options with a security consultant intelligently.

We know from years in the business that wireless security systems provide the best security that any system has to offer, and we admit from the outset that we hope after reading our book, you will buy a home security system. Put a wireless system to work along with a comprehensive life-safety plan and your chances of becoming a victim of crime will be significantly reduced.

With this book, we hope to save you the headaches that come with poorly designed security systems, careless installations, and uninformed salespeople who might be more interested in selling you a gadget than providing you with the information you need to make the right choice for the security needs of you and your family.

# Security and the twenty-first century

It's a fact of life right now in the security industry that most security systems are sold, not purchased. That is, most alarm systems are sold by independent dealers and national chains by phone, via direct-mail campaigns, and by referrals. Only a small number of people decide one day they want a security system in their home and then shop for the components at an electronics store the way they would a VCR, TV, or CD player.

A few do-it-yourself kits are available at chain stores, and, of course, some people do shop around and pick one up off the shelf, return home, and install the system. But most people buy alarm systems from dealers whose business it is to sell them.



So why don't people go out and purchase an alarm system for their homes? One reason is that, unlike people who go out and buy a CD player because they know they want to listen to music, most people are not absolutely certain they are going to be targeted for a burglary. People who invest $3,000 in a home theater system know they'll enjoy watching movies in their family room, but nobody knows whether their house is going to catch on fire. So it's really no mystery that alarms are actively sold by alarm companies, instead of purchased by consumers.

What is a mystery is why so many people are willing to spend tens—even hundreds—of thousands of dollars on household furnishings, electronics, art, sporting goods, and collectibles, and not spend a fraction of that amount on a proven deterrent to theft and an aid to life safety—an electronic alarm system.

Lou likes to tell the following story of what it's been like selling alarm systems in the last 20 years. The story will aid in an understanding of the importance of security systems as we head into the next century.

## One salesperson's story

Twenty years ago I was a new sales manager with a fire alarm franchise in the New Orleans area. The day I started work, the owner of the franchise directed me to the offices across town. To get a better feel for the company and for the product I would be selling, I started calling existing customers.

I had a list of questions prepared to ask each customer. My goal was to find out why they purchased an alarm in the first place and whether "buyer's remorse" existed in the alarm industry. Much to my surprise, almost everyone I called was happy with their purchase and would buy again from us. In fact, I even found some customers who wanted additional sensors. I also learned that most bought their system after a problem had occurred in their neighborhood, at their own home, or at the home of a friend or relative.

After many calls, I was speaking with a woman who had positive comments about everything until I asked, "What did you think about the salesperson?" To that question she answered, "You want to know what I thought of the salesman?" Based on the tone of her voice, I knew I had finally found the irate customer. So I lied. I said, "Yes, I would."



She continued, "He was the pushiest salesman I've ever met. He wouldn't take no for an answer. I kept telling him I wasn't interested in a fire alarm system. My husband told him we weren't interested. But he just kept on talking."

"I'm very sorry," I said. "These things happen occasionally, but I apologize if we upset you."

"I'm not finished yet," she said. "I finally got so tired of him that I just got up and went upstairs to bed. I left my husband and your salesman downstairs. The next morning, I asked my husband what happened. He said that he bought the fire alarm system just to get rid of the salesman. He was convinced that he wouldn't leave otherwise."

Once again I jumped into the conversation and apologized for the pushy salesman. But once again she told me she was not finished yet.

"This all happened just before the Christmas holidays," she explained. "Every year, just before Christmas, we go to a neighbor's house for a little holiday get-together. We leave our children, who were 10, 12, and 13 years old, at home alone. We always felt okay leaving them alone at the age they were."

"We had been at the party for about two and a half hours when we heard a commotion outside. We heard sirens and scrambled out the door. I looked down the street and my heart jumped into my mouth. My house was on fire. I ran toward my house, thinking, *Oh no, oh no—not the kids!* I had always thought that nothing could be more horrible than dying in a fire, and I couldn't bear the idea that my kids were inside the house."

"When I arrived, to my joy, my three kids and a fireman were standing on the front lawn, safe. The fireman asked if I had installed a fire-alarm system. I said I had. The fireman said, 'It's a good thing you did because based on the speed the fire spread, without the early warning of a fire-alarm system, your three kids would surely have died.'"

"You want to know what I think of your salesman?" she asked me again. "I owe my kids lives to his belief and conviction that I needed a fire alarm system. I pray and give thanks for that salesman every day of my life."

Now you know why we're writing this book and why we want you to install a security system. But why a wireless system? For one simple reason: Although they are the most misunderstood products in the security industry, we believe they're the best. They're the most reliable, the most sophisticated, the most convenient, the easiest to install, and the most cost-efficient security available.



Because fire alarms and burglar alarms are life safety systems, this is one profession that contributes to the well-being of its clients. Security systems save lives. And wireless security systems do it with some advantages and features that other systems can't match.

## What's ahead in security system technology

The twenty-first century will benefit from changes that are already taking place and are being perfected. Some wireless security systems are already designed to fit into integrated home-management systems for controlling lights, heat/AC, appliances, as well as security and access control. (*Access control* refers to the means of controlling who gains access to buildings or to rooms inside buildings.)

Perhaps the greatest change underway is how and where wireless systems will be available for purchase. Building contractors are already starting to include home security systems—or at least the basic components and prewiring for power and phone lines—in their new-home construction plans. Utilities are beginning to package security with their automatic meter reading technology, because both technologies use digital phone line communications to transfer data from the home to a monitoring station.

Companies that now sell systems only through dealer networks will no doubt make simpler, less sophisticated versions of their more complex systems and package them for sale in home and hardware stores for do-it-yourselfers. Companies have already emerged with systems designed exclusively for the DIY market. (As we'll discuss in later chapters, wireless is perfect for do-it-yourselfers because few tools and no wiring are required. But because of the wide range of difference among systems, you'll want to carefully weigh the pluses and minuses of DIY packages versus professionally installed systems and whether to have a monitored systems or a simple alarm.)

Wireless technology is indeed in its golden age, and the situation has never been better for consumers. Prices are falling, not rising. Systems are doing more, not less. And consumers have more choices than ever.

## In the next chapter



In this chapter, you learned of a husband and wife who bought a fire alarm system to get rid of a badgering salesman—not because they thought something could happen to their family. In Chapter 2, we'll look at a psychological factor called the "law of self-exception," which prevents people from believing they could be the next victim of a fire or crime and allows them to drop their guard. We'll also look at ways you can improve the safety of your family and the security of your belongings that do not involve installing a security system.

# Determining your SQ

2

Before going to the effort and expense of securing a home, a person has to believe that the threat of burglary and fire is real. Apparently, many people do not believe they are likely to be the victim of a burglary, since the second-most common response to the question, "Why don't you own an alarm system?" is that the person never thought about it. To date, only 16.5 percent of U.S. homes (approximately 16.5 million) are reported to have burglar alarm systems. That leaves a lot of unprotected homes. This chapter will help you sharpen your SQ (security quotient) by asking you to apply common sense and focus on your vulnerabilities.

## The law of self-exception



We are all ruled by the psychological law of self-exception. In short, this law tries to convince us of our immunity to all kinds of dreaded diseases and harm. The smoker living under the law of self-exception continues to smoke two packs of cigarettes a day despite the abundance of research and warnings that smoking causes emphysema and lung cancer. The law convinces the smoker that "it won't happen to me."

Personal security and the protection of our loved ones from the twin perils of burglary and fire—and most recently the third danger posed by carbon monoxide poisoning—is something to take very seriously. However, before anyone can do that, he or she must first feel that the threats are real. There's no way of knowing how many of the 2,670,000 burglary victims in 1994 thought the threat was real, or how many residents of the 314,000 one- and two-family homes thought that their houses would catch fire. The psychological law of self-exception allows people to drop their guard in the belief they won't be the next victim.

Following are a few typical examples of this law:

**"Burglars are not likely to break into my home because I have nothing to steal."**

Nothing but radios, computers, cameras, TVs, and VCRs. Burglars break into homes of all kinds, looking for all kinds of things. In most cases, burglars are looking for anything that can be fenced. The more common the item, the better. Sure, you wouldn't sell your old TV for the 50 or 100 dollars you could get for it, but a burglar would gladly take that much for an item he didn't have to pay for. Burglars even break into boarded-up, empty homes. They steal the flooring, doors, bathroom fixtures, copper wiring in the walls, and anything else they can remove. All of these things have a value and can be exchanged for cash.

When the law of self-exception is lifted, you start saying, "I'm just as vulnerable as everyone else. There's no way to say with certainty that I won't be arbitrarily targeted for a burglary." When you know you are as susceptible as others, you start making your home less attractive to burglars.

**"I live in a nice neighborhood."**



Nice neighborhoods are just as much a target as any area you would consider not nice. Statistics suggest most burglars live within 3 miles of their victims. Contrary to popular belief, burglars are not people who only live in the rougher areas of town. The burglar might live on your street, next door, or around the corner.

**"I live in a gated community."**

If the burglar looked like a stereotype of a burglar, a guard stationed at the entrance of a gated community would recognize him and keep him out. However, because a burglar can look like anyone else, and because burglars learn to buffalo their way past security guards and police, you are a potential victim whether you live in a gated community or not.

Fences do not keep dishonest people out; they keep honest people out. Fences provide privacy for you and the burglar.

**"We have good, nosy neighbors. They'll see a burglar and call police."**

The best-intentioned neighbors cannot ensure that a burglar won't break into your home. Several years ago my next-door neighbor was burglarized while my wife and I ate dinner at the breakfast table not 30 feet from the door the burglars kicked in. We didn't hear a thing.

Like everyone else, our air conditioner was running, a radio or TV was playing, and we might as well have been on another planet. Sure, we would have called police if we saw a stranger lurking around our neighbor's home. But we didn't see him.

We have spoken with burglary victims whose neighbors saw someone near their home but didn't think the person was a burglar. With our inclination to not get involved, with the tendency for us to live in neighborhoods and not know the family who lives in the house three doors down or even next door, it is not likely we will call police when seeing a stranger in a neighbor's backyard. Our neighbor is a stranger.

**"The police or a neighborhood patrol patrols the neighborhood."**

A police patrol or a private security patrol certainly lessens the risk of street violence and, to a lesser degree, burglary. But unless police drive up as a burglary is in process, assuming the burglar is burglarizing a home from a visible spot, police won't know the burglary is happening. Police patrol is effective in combination with a monitored security alarm system. Short of that, you are still at risk.

In a study of the community of Greenwich, Connecticut, Dr. Simon Hakim of Temple University found that in 74.3% of unsuccessful burglary attempts, an alarm deterred the burglar. Based on his extensive research on crime and the effect of alarm systems, Dr. Hakim has noted that the chance of burglary for unalarmed homes is 2.7 to 3.5 times higher than that of alarmed homes. So the question is, which 2.5 million homes will be next year's burglary statistics?

Along with the absence of an alarm system, certain innocent habits can increase your chances of becoming a burglary victim. Unavoidably, we display our daily routines for all to see. The kids get on the bus about the same time every day, we head for our jobs at about the same time, and we arrive back home on a fairly predictable schedule. It's obvious that millions of homes are vacant between 9 AM and 5 PM. A professional burglar need only park on your street for a couple of mornings to learn the habits of your family and those of your neighbors. Determining when it is safe to break in is fairly easy.

## The good news and bad news in crime and fire statistics

Since 1988, there has been a downward trend in residential fire deaths, including a 7.8% drop in deaths between 1993 and 1994.

This trend is partly due to the installation of smoke detectors. In 1977, approximately 22% of homes had smoke detectors and there were 5,865 fire fatalities. By 1993 approximately 92% of homes has smoke detectors, and the number of deaths had fallen to 3,720. That's the good news.

The bad news is that, according to the National Fire Protection Association, 614,000 structure fires occurred in the United States. Of these, 341,000 were one- and two-family homes, 97,000 were apartments, and 13,000 were dormitories, boarding houses, and hotels/motels. The South and the Northeast have the highest fire incident rates, with 9.3 fires and 9.0 fires per 1,000 population, respectively.

Burglaries are on a downward trend, too, although, as mentioned, there were over 2.58 million of them in 1994. Robberies fell for the third straight year in 1994, down to 627,000.

Burglary rates are highest in the months when people are most likely to be on vacations away from home. According to studies conducted by Dr. Hakim, burglars tend to strike most often in July and August. In addition, burglaries tend to occur along major arterial roads and fall off with greater distance from freeway exits. Also, nearly 34 percent of burglaries occurred in homes occupied by their owners for fewer than five years. Such homes are often more vulnerable to burglary simply because the owners have not yet gotten around to taking security precautions.



Keep in mind that burglars predominantly hope to enter a house quickly without being noticed. In one study, 82 percent of entries were on the first floor, with the back door the more popular point of entry. It is interesting to note that 12 percent of burglaries are through an unlocked door.

People with incomes between $50,000 and $150,000 experience more burglaries than owners with higher incomes. Unalarmed homes whose owners make between $50,000 and $100,000 a year have nearly twice the burglary risk of alarmed homes with owners in a similar income range, according to one of Dr. Hakim's studies.

The average value of property stolen goes up in unalarmed homes. The higher amount can be attributed to the longer time a burglar has to stay on the premises.

Statistics can help convince us to take steps to avoid becoming the next victim. Taking several steps—having a dog, installing security and fire-alarm systems, and using interior and exterior lighting, alarm warning signs, window decals, and deadbolt locks—will help keep the good news coming and the number of fire deaths and burglaries falling.

## What you can do to fight back

Many of the steps you can take do not add expense to an alarm system. For instance, to avoid predictable daily patterns, you might want to vary your schedule, but doing so can be a complicated matter. The better solution is to implement a combination of security measures that will keep a burglar from considering your home as a target, that will deny or delay entry if you become a target, and that will detect an intrusion at any point of entry to your home and sound the alarm.

To keep some burglars from even considering your home, place an alarm company yard sign out front and stick decals on all windows. The sign and decals are enough to scare off burglars who are looking for the easiest place to break into.

However, if the burglar is especially determined or desperate, the yard sign and window decals might not work. In this situation, your defense strategy is to delay entry as long as possible with good locks. If you are inside, you can escape and notify authorities. If you are not inside, the burglar will simply have to work harder, while the chances increase that he will be spotted by a neighbor.

Although a lock's function is to deny access to unauthorized persons by providing a physical barrier to the intruder, they can be defeated by a determined burglar. A good kick is all it takes to break some doorjambs, and a hard yank on a crowbar can do the job when a kick doesn't work. It's at this point that a detection system does its job by signaling an unauthorized entry, setting off a local siren, and communicating the intrusion to a central monitoring station.

Even for successful burglaries, an alarm system still presents an advantage. According to the Hakim study, burglars who persist in spite of all security measures know that they have far less time to scour the premises for valuables and on average leave with 31% less than they do in unalarmed homes.

## Home security while you're away

It's not that hard to tell who is on vacation and who isn't. If no garbage is out on pickup day, it's likely nobody is home. Instead of allowing this piece of evidence to become an invitation to a burglar, ask your neighbor to put part of their garbage in front of your home. Offer to do the same for them when they are away.

## Newspaper delivery

In addition, many security experts recommend that you instruct the newspaper office to stop delivery during a vacation or extended trip to prevent a buildup of newspapers. For the most part, this is good advice. However, in one case the plan backfired:

In the mid-1970s, a rash of burglaries broke out one summer in a community near New Orleans. Every burglary was perpetrated against a homeowner who was on vacation when the burglary occurred. During the investigation, police interviewed each victim and discovered the only common denominator was that each had stopped the newspaper prior to leaving on vacation. They all had given the newspaper office notice of when they were leaving and when they expected to return home. The obvious conclusion was that the newspaper carrier was the culprit.

Upon further investigation, however, the police determined that the victims didn't all have the same paper carrier. The thought that several newspaper carriers could be involved was beyond belief. Surely this couldn't be the work of a theft ring. After all, we were talking about paperboys here! But the commonality of the burglaries couldn't be ignored, so the police set up a sting operation to see if their suspicions had merit.

Several calls were made by police posing as newspaper subscribers. They told the paper carriers they would be out of town for two weeks and for them to suspend delivery of the paper until they returned. During that time, undercover police officers staked out the homes and waited. The result of the stake-out was the arrest of a half dozen paperboys who were, in fact, burglarizing the victims' homes.

The newspaper routinely posted requests to stop delivery on the bulletin board so the paper carrier would know not to deliver. Anyone who wanted to could simply read the notices to learn who is out of town and for how long. Interestingly, the paperboys had not limited their targets to homes that were on their routes.

So for added security when you're away, don't call the paper with a suspend order. Ask a trusted neighbor to pick up the papers for you. A lot fewer people will know of your travel plans if you do.

Also ask your neighbor to remove flyers that make their way onto your doorknob or stair railing.

## Mail delivery

Similarly, it's not a good idea to allow mail to accumulate in your mailbox too long. As with the newspaper, ask a neighbor to pick up your mail. However, if that can't be done, you can arrange to suspend mail delivery for a specific period of time. Keep in mind, however, that you are subject to the same potential problems that may come with stopping the newspaper. Again, the fewer people who know your plans, the better.

## Grounds maintenance

Grass cutting, along with basic sidewalk, driveway, and grounds maintenance/cleaning, should be maintained in the same, normal manner as when you're at home. The key word here is *normal*. We know of one man who advertised his absence in the most unusual of ways. He was not known for his expertise in lawn care. His grass was cut all too infrequently, grass clippings were usually left where they fell, and it wasn't unusual for him to leave parts of his lawn maintenance equipment at the exact spot where he finished.



Most every summer he drove off with his family to a lake home for several months. When he was at the summer cottage, necessity coupled with the loud protests from neighbors demanded that he contract with a lawn care service to make the yard presentable during the family's absence. To anyone who looked, the difference between when the family was in town and when they were away was like night and day, and any perceptive burglar would know when the time to strike was right.

## Loose lips sink ships

Don't broadcast your travel plans to anyone who doesn't have to know. This includes your hairstylist, mechanic, and bank teller. The threat of a break-in could come from an innocent remark overheard by the wrong person. For example, say you're at the repair shop and you tell the mechanic you want your car road-ready for next week's trip. Who in the waiting room is listening? A burglar can take a quick glance at a repair order to get your address and then wait for signs that you've left town.

Or suppose a bank teller selling you traveler's checks starts a friendly discussion about vacations, and you say you never thought you'd be spending a week in Greece. Who is watching you sign your name in order to match it with an address in the phone book? Hair salons and restaurants are also favorite places for criminals to pick up on your travel plans. You'll go a long way toward preventing a break-in by discussing your travel plans on a need-to-know basis only.

### Interior and exterior lighting

Someone observing your house for signs of vacancy will look for the obvious differences between the lighting in your home when it is occupied and when it is not. With some variations, you turn off lights in the living room, then the bathroom, then the bedroom. Some people think you can trick burglars into thinking you're home by turning an outside light on and keeping it on for the duration of your time away. Others put all the lights on timers so they all come on at the same time and all go off at the same time. Keep in mind that it's much wiser to stagger the ON times and the OFF times to create the illusion of someone walking through the house turning off the lights.

In order to minimize your vulnerability, electronically duplicate the look your home takes on when you are at home. Lights, TV sets, and radios all can be programmed to turn on and off according to a preprogrammed schedule. Your home will look alive and therefore less appealing to the burglar. At first, these techniques might seem to stem from paranoia; however, once you believe that you could be the target of a burglary, certain actions take on the air of common sense. Burglars use all the tools that are available to them, and it is your job to create an environment that is not appealing to them.

Leaving lights on all the time when you are away actually works against you. Do you leave a porch light on all day and night when you're home? Probably not. Do you leave an inside light on all the time? Or a TV? If you leave a porch light or an inside light on continuously, as so many people do, your absence become obvious.

The computer and electronics industry provides the tools necessary to create that "I'm at home" look. Some wireless security systems also do it, with the added bonus of summoning police help should the burglar break in anyway. The solution is only a fraction of the cost of a burglary, and is insignificant when compared to the grief one suffers after the privacy and protection of one's home has been violated.

## What's your SQ?

Being smart about security is sometimes a matter of common sense. Although good locks, a high-quality security system, proper lighting, and a barking dog are all effective components of a home security plan, each has a different function. Using a variety of security measures is the best protection against losses due to break-ins. But even if you use them all, sometimes you make yourself vulnerable to burglary when you don't take just a moment to do the smart thing. Here are 19 simple home security tips that will improve your SQ and make your home a safer place.

1. Study your own personal habits. Decide for yourself if you could be advertising your absence from home. Even when you are not on vacation, consider how you might be telegraphing an empty home or, just as bad, a home where only you or one of your children are home alone.
2. Never allow newspapers or flyers to clutter your doorway or walkway. Make arrangements for someone, preferably a neighbor you can reciprocate with, to pick up and discard newspapers and flyers.
3. Ask the same neighbor to place a garbage can out for you on pickup days and to remove the empty can after pickup as you would normally do.
4. Ask a friend or neighbor to park a car in your driveway if it is normal for a car to be parked there at night or during the day.
5. Don't allow mail to accumulate in your mailbox. You can arrange for a temporary stop at the post office; however, having a neighbor pick up your mail is preferable. The fewer people who know you're away, the better.
6. Have your grass and gardens cut and trimmed while you're away. Don't have those services done in a manner that isn't normal for your home. Remember, if it looks better than normal, you've sent a signal to the very observant criminal.
7. Don't turn a porch light or any other light on for the entire time you are away. A light burning during the day is a dead giveaway. Use timers or a security/home automation system that causes lights, TVs, and other appliances to mirror your habits as though you were home.

8. A key in the mailbox, under a mat, or over a door is a key within a burglar's reach. Take a minute to make other arrangements for getting keys to authorized persons, such as leaving a key with a trusted neighbor.
9. Are you in the habit of leaving your keys in the door lock? Unless you break the habit, you're inviting a break-in.
10. You say you're just "running to the store for a few things," so you don't bother to close the garage door. A car-less garage is a very good sign that no one is home. Keep security a priority, and you will prevent the loss of bikes, lawn mowers, and power tools, which take but seconds to steal.
11. Many locks are easily opened by a credit card, knife blade, or even a slat from a venetian blind. Add deadbolt locks with bolts at least 1 inch long to exterior doors that don't have them.
12. Unlocked windows are the burglar's second-favorite point of entry. Take a few seconds to fasten locks. If you have a security system, a simple thing like a window warning decal can also discourage a burglar.
13. Ladders make second-story work a breeze. Don't leave burglars this engraved invitation. Instead, make one more trip to the garage and put the ladder safely away.
14. A dark house provides good working conditions for burglars. Leave lights on to make your home an unpleasant place for a thief to work.
15. Burglars prefer side, rear, and garage doors because they are often shielded from view of the neighbors and the street. Use high-quality locks for these low-profile points of entry and keep them well lighted.
16. A note on the door that says you'll be gone for half an hour gives a burglar a 30-minute plan of action, and a telephone message that says you'll be gone for the weekend could spell further disaster.
17. Valuables visible through the window take the guesswork out of which house to hit. Display only curtains and blinds in your windows.
18. Many burglars have day jobs. Be careful about handing over keys or discussing vacation or evening plans with service providers like bank tellers, cashiers, and others.

They may engage you in casual conversation so they can casually haul off your belongings.

19. Keys are an important part of your security plan. They are also easy to copy. Don't hand over your home to valets or parking lot attendants. Keep car keys separate from house keys when leaving your car with strangers.

## Telephone security

The telephone can a good friend; it can be a comfort for the lonely and a lifeline for the infirm. But it can also be our worst enemy if not used properly. Advances in the features provided by the telephone company have reduced some of the problems, but many still exist.

The key to telephone security is asking questions and not giving answers. If someone calls and asks for a person who doesn't live in your home or isn't someone you recognize, simply advise the caller they've called a wrong number, then hang up. If they call back asking for the unknown person again, ask what number they are calling. If it is your number, advise them again that no one by that name lives there and hang up. If it is not your number, tell them they have misdialed, but don't give out your number.

People will disclose a great deal of information over the phone to a stranger. This tendency stems from a natural impulse to be polite. Con artists rely on that impulse and almost always speak with an air of authority. They know how to ask for and get sensitive information from you over the phone. However, if you follow the above rule, you can't lose. Don't give information of any kind by phone to someone you don't know. Don't answer any questions regarding your security, number of people in the home, financial data, or anything else you wouldn't disclose on page one of the newspaper.

Let's emphasize this point by looking at the difference between what you might say over the phone and what a burglar who is "casing" your house actually hears:

*You said: "I can't talk now, I'm on my way out."*

You meant: "I'm too busy to talk."

*The burglar heard: "I'm leaving the house empty."*

You said: "He's/she's not here right now."

*You meant: Your spouse is not available to come to the phone.*

The burglar heard: "I'm home alone."

*You said: "This is 555-4455."*

You meant: "You didn't call the right number; you called my number."

*The burglar heard: "I am home." (In this case, the burglar is simply able to confirm the presence or absence of a person in the house that is being targeted.)*

You said: "I'm the baby-sitter."

*You meant: "I don't know the answer to your question," or "I'm not the right person to ask."*

The burglar heard: "Just the kids and I (usually a very young person) are home."

## Answering machines

While we're on the subject of telephone security, let's consider how you program your answering machine. Don't record messages that tell strangers you're not home or are out of town. Sometimes machines will give a time when the person expects to return, which is not a good idea if a burglar is on the other end of the line. All the answering machine has to say is that you are away from the phone at the moment. This message would be the truth whether you are in the shower, in the yard, or in the south of France.

A final point on answering machines: Check and clear messages often, even if you're out of town. The caller can often determine by the length of the tone that a lot of messages have been recorded, possibly giving away that you are away from your home.

## Harassing phone calls

Harassing or obscene phone calls can be very upsetting. If you receive such a call, be sure to stay calm. The caller might want you to respond emotionally, and if you do, you might be encouraging the caller to call back. Do not say anything to the caller. Instead, calmly hang up the phone. Even if the caller never speaks, such a call can be considered harassing. To discourage the caller, blow a whistle into the receiver. Keep track of dates and times of calls and report them to the phone company. You might also consider investing in a phone with Caller ID, which displays the number of the caller to you.

## Security and latchkey children

Children who are home with no adult supervision need to know what to say when answering the phone. With so many working parents, burglars know children are frequently left alone for certain periods of the day. They also know they can often con children more easily than adults.

Teach your children telephone security as readily as you would teach them about the dangers of fire, chemicals around the house, and appliances. A child who doesn't know the caller should never say he or she is home alone. If asked about the whereabouts of Mom or Dad, the correct response is, "They can't come to the phone. May I take a message?"

If the caller persists and asks when the parent can come to the phone, a safe response is, "I'm not sure. I'll pass along the message you called. Who is calling?" If the caller persists, instruct the child to say "good-bye" and hang up. Burglars will try to drill a child for information to determine if the child is alone. Children should not answer any questions asked by a stranger.

Burglars will pretend to be from the electric company or gas company, and talk their way into a home with a fabricated story of a gas leak. A stern, authoritative person demanding entrance will often coerce a child and gain entrance into the home by sheer intimidation. If the doorbell rings, kids should check to see who is at the door without opening it. A wide-angle peephole at the height a child can see through is helpful here. An intercom system is good, too. The child should ask who it is and what they want. Children who are not expecting anyone should not let anyone in.

Encourage children to stay inside until the first parent arrives. Keep sufficient snacks and entertainment for kids to enjoy during the time when children will be unsupervised. Encourage kids to call Mom or Dad or the names on an emergency phone list if they are not sure what to do in any situation.

Some alarm systems offer a "latchkey" feature that calls in an alarm unless the call is canceled by the child within a specific time frame. The alarm monitoring center will have all of the appropriate phone numbers to reach the child's parents, along with those of grandparents and other relatives or friends. All this happens automatically if the child doesn't cancel the call on time. If the child does, no calls are made.

## Parenting for life safety

Keeping a home safe and secure, and teaching kids habits that can prevent harm are among the greatest gifts parents can give to children. Keep the following tips in mind to help prevent dangerous situations:

- ☐ Leaving children alone for just a few minutes at home, in the car, outside, or in a store makes children vulnerable to accidental and intentional harm. Traffic accidents and falls take place in mere moments. A person intent on abducting a child needs very little time to act. Leave your children only with trusted friends and relatives.
- ☐ Only hire baby-sitters whose references you have checked. Ask prospective baby-sitters for the names and phone numbers of two or three persons for whom they have worked who aren't their relatives. Ask the references for specific strengths and weaknesses about the sitter. Explain the ages and temperaments of your children and ask how well the sitter would fit with your kids.
- ☐ Train your baby-sitter in the use of your home security system. Establish rules about answering the phone and the door. In the event of a medical emergency, establish whether the sitter should call you first or a doctor.
- ☐ Point out the location of fire extinguishers. Explain that in the event of a fire, you expect a sitter to get everyone out of the house before calling the fire department from a neighbor's home. Emphasize that once everyone is outside, no one is to reenter a burning house or apartment building for anything.
- ☐ Always be sure to leave a number where you can be reached in an emergency. Also leave the name and number of neighbors or relatives who live close by.
- ☐ Instruct your children in a few life safety rules. Agree that children are to come straight home every day from school or from after-school activities. Tell children they must notify you immediately if there is to be a change in after-school routine. Children should avoid talking to strangers and *never* get into a car with someone they don't know. Remind children that they do not have to be polite to strangers who get too close or who try to get them to go for a ride.
- ☐ Hold a practice session in which you instruct children to run away if they feel uncomfortable with a stranger. The parent plays the role of the stranger and says something like, "Why

don't I give you a ride—I'm going that way anyway." And the child must practice running away, even if it sounds like a friendly invitation.

## Preventing a carjacking

The FBI describes carjacking as auto theft that includes a crime against a person in the taking of the vehicle. Physical force is always a possibility in the commission of the carjacking, but injury or death can also result if the victim is dragged or struck by the car.

Carjacking appears to be a crime of opportunity, so keeping opportunities to a minimum decreases your chances of being a victim. Park in well-lighted areas and always lock your car. Weigh the desire for inexpensive or free parking with the increased chances of being targeted on a dark, uninhabited street. As you approach your car, look for signs of trouble such as strangers watching you or hanging out near your car. Have keys ready and get into your car quickly. Lock the doors, start the car, and go. A parked car is no place to count your money or put on makeup. Keep the doors locked and the windows up while driving.

According to the FBI, the majority of carjacked vehicles are recovered. Many vehicles are stolen as a means of getting away after robbing the driver of valuables. Therefore, it doesn't make sense to try to resist a thief who is holding you up. You can replace almost anything the thief takes—except your life. Resist the temptation to fight back or to express your rage at the thief, whose desperation is impossible to gauge. Instead, make a plan of least resistance ahead of time and stick to it if you're ever held up for your car.

Unfortunately, you're not much safer in an older-model car than in a new luxury model. The FBI reports that carjacking occurs across the spectrum of year, make, and model. Parking lots are the favorite areas for carjackers and so are areas around automatic teller machines. The reasons are obvious. You have the most cash when you're shopping or when you've made a withdrawal. Take extra precautions under these circumstances.

## Use touchpads, not handguns, to protect your family

It has been disturbing to hear about the tragic deaths of children who have been killed by their own fathers, who suspected them to be intruders. A newspaper story of a Texas father who killed his daughter when he thought she was a burglar is only one of several reports from the 1990s. Some simple steps can be taken to avoid such a tragic outcome:

- ☐ Once a burglar has gained entrance to a home, the safest response is to escape and have proper authorities take over. Even if an intruder should get away with jewelry, some electronic gadgetry, or a handful of cash, that is a far better outcome than personal injury or death. The security industry designs detection systems to prevent loss of life and property. When used properly, these devices save lives. A detection system at the home of the Texas family did what it was designed to do and decreased the potential for harm. When the father went into the house armed and looking for a burglar, he decreased the life safety benefits of his security system.
- ☐ Remember that the vast majority of burglars do not want a confrontation. If you come upon an intruder in your home, do not place yourself between the intruder and an exit door. Let the intruder get away. If the intruder won't leave, you should leave and notify the police.
- ☐ If you are outside your home and you believe an intruder is in your home, stay out of the house and notify the police.
- ☐ Mistakes are very easy to make with handguns. You might miss the intruder and hit an innocent person. The intruder might get the gun from you, panic, and shoot.



Police procedures vary, but authorities recommend the following precautions to avoid being hurt or hurting an innocent person when you are armed and suspect that an intruder is on the premises:

1. Take cover.
2. Identify yourself.
3. Tell the intruder what you want them to do, for instance, "Don't move. Put your hands up."
4. Contact the police immediately.

It is frightening to have an intruder in your home. But untrained persons who try to apprehend criminals do so at great risk to themselves and possibly to others. Following these steps can greatly reduce the chance of a tragic mistake.

## What a difference a door makes

Putting a lock on a door seems like good security, but neither doors nor locks are created equal. Good security requires that your lock *and* your door be of high quality. If one is weak, the other becomes weak. A good lock on a cheap door provides little in the way of protection. A hefty shove is all it takes to break through that kind of security. Checking the security of your exterior doors takes just a few simple steps.

A front door that provides good protection against a break-in is one that provides a stable, secure home for a good lock. The door has to be made of a material strong enough to resist being pried, hacked, or hammered. The most secure doors have a solid hardwood core of oak or other hardwood. The locking mechanism cannot be easily broken out of a hardwood door. The door should be hung in a hardwood frame that resists spreading. Doors should fit closely and swing easily on strong hinges. Although slightly less secure, doors with interior cores of particleboard or softwood are still very secure and cost less than the more expensive hardwood varieties. Hollow-core doors, commonly found on interior doors, make ineffective front doors because they don't provide much support for even a strong lock.

Take a look at your front and back doors and notice the locking mechanisms. Many doors have deadbolts that extend from the edge of the door into a strike, or metal plate, secured to the door frame. The deadbolt should extend an inch or more from the edge of the door. The strike should be secured to the doorjamb with screws that are 3 inches or longer so they secure the plate to the framing studs and not just the doorjamb.

Some homes have doors with an entrance handle set that includes a deadbolt lock located on the door above a doorknob or latching mechanism, which may or may not have a lock in it. A high-quality deadbolt of this kind can provide adequate protection. However, if a key is required to unlock the door from the inside of the house, you must weigh the security advantages of the cylinder deadbolt with an important fire-safety concern: there will be little time to look for the front-door keys in the event of a fire.

If you find flaws in your doors, you don't necessarily have to replace them. There are ways to reinforce doors and doorways to make them more secure. But locks that are cheap, worn out, or inoperable should be replaced. When you consider their value in securing your home, it's money well spent.

Following is a list of questions to gauge the security of your doors:

1. Open your front door and look out: Is there anything about what you see that makes your front doorway more vulnerable than other areas? Is there a clear, unobstructed view of your door from across the street? If someone were standing at

your front door at night for a few minutes, would they be noticed easily? How well do you know your neighbors? Would anyone pay attention to the person standing there? Would anyone call you to advise you of the person standing there?

2. What kind of lock do you have on the front door? Does it have a deadbolt? Is it keyed on both sides?
3. Is there any glass in the immediate vicinity of the front door? How big? Could a man get in through the glass? (Remember, if a burglar's head can fit through an opening, a burglar can squeeze his body through the same opening.) If you have a thumb bolt on the inside of your door instead of a key lock, can the thumb bolt be reached from the glass?
4. What is the construction of the door? Is it solid wood, glass and wood, or metal? Decorator wooden doors with nice carved panels are a piece of cake to break into. The wood at the edge of the decorator carving is only about ⅛ inch thick. With a few well-placed kicks, the burglar can break a hole in the door.
5. What does the jamb of the door look like? Is it wood? How long is the throw of the dead lock? How is the keeper (the female side of the lock installed on the jamb) installed? Does it have long (secure) or short (not secure) screws?



Burglars break into front doors a number of different ways. One way is for the burglar to kick the door at the location of the lock. In most cases, the door will fly open, tearing away the wood of the doorjamb. The higher on the door the deadbolt lock is installed, the harder it is to kick the door open.

Another way in is to use a pry bar in between the door and the jamb to pull the bolt far enough away from the jamb to cause the door to open. Cheap door locks have short throws and can be defeated easily with this method. Even long deadbolt throws can be defeated through the use of a body-and-fender shop jack. This jack is placed between the opposing doorjambs. The burglar then jacks the frame of the door until it bows and the door swings open.

## False alarms

We talk about false-alarm prevention in detail in Chapter 17, but it's important to address the issue here briefly, since many people are apprehensive of installing a security system because of them. And for good reason. Many cities have instituted fines for false alarms. In fact, there are 24 million false alarms each year. That's 94 to 98% of alarm activations! In Philadelphia alone there were 151,795 false alarms in 1994. At an average cost of $28 per response, the city of Philadelphia spent $4.3 million on false alarms.

In response to the problem, Seattle enacted an ordinance that lets police officers leave a $50 citation at the scene of the false alarm. The measure has been effective, proving that many false alarms are preventable. The ordinance resulted in a 14% drop in false alarms. Plans like the one set up in Seattle work well because users are responsible for 75% of false alarms, and the errors are preventable. Practice and a little care are all it takes to prevent false alarms.

Another very effective preventative measure is to include a central monitoring station in your security plan. Monitoring stations are a big help in reducing false alarms. A monitoring station generally screens out as many as 90% of the alarms actually received and only dispatches police on the 10% or less they cannot verify. When an alarm is set off, the monitoring station can call the home before any units are dispatched. In the event of a false alarm, the system user then provides a user code and states that the alarm was accidental. Monitoring stations do not give fines for false alarms.



Many false alarms are caused by businesses. Employees who are responsible for closing up at night and turning on the alarm system can be careless in doing so. The more turnover a business has, the greater number of false alarms they create due to untrained users.

Complacency on the alarm companies' part is also a factor in false alarms, but once cities stepped in with threats of fines that would impact the business of selling alarm systems, the alarm manufacturers got serious, and some have invented systems that eliminate equipment-caused false alarms.

Many people choose not to buy alarm systems on the basis of the false alarm issue. That's like deciding not to drive a car because you might run out of gas. With planning, you can easily prevent being stranded in a car with no fuel; you can also take measures to prevent false alarms:

1. Expect thorough training from your alarm dealer.
2. Stay away from cheap equipment whose electronic components are susceptible to false alarms.
3. Train all family members in the use of your system.

4. Have your system monitored by a professional monitoring station: Your monitoring station can prevent 90% of accidental alarms from resulting in false dispatches.
5. Only buy a system from an alarm company that is willing to provide excellent customer service.

## In the next chapter

Now that you understand why a security system is so important, we'll look at why wireless systems in particular are such a good investment. Before we do that, however, we'll discuss the drawbacks of the technology of the past. We have one primary reason for doing so: to correct the many misconceptions and the misinformation still surrounding wireless systems so that you can make informed decisions as a consumer.

# 3 Wireless security: The past

In the 1970s, easily 98% of all professional security alarm companies installed only hardwired alarm systems. The majority of systems sold for residences were for fire. Occasionally, a panic button would be added to the fire system, but few burglar alarms were sold. Yet even in the early 1970s, some of the systems installed were wireless.

Prices were extremely high in the 1970s and on into the 1980s. Hardwire security was hard to install, and it was costly. The only customers who could afford to buy these systems lived in enormous homes. The average installation with at least two very well-trained installers required three or more installation days, and on many occasions, more than a week.



Finding and training professional alarm installers was another problem. The job of installing a hardwire alarm system had less to do with one's electronic acumen than with one's ability to fish wires through basements, floors, walls, and attics. Because the job of installing hardwire security was so highly specialized, talented, trained installers were hard to come by. Once found, a reliable employee interested in the alarm business would take up to two years of apprenticeship before being considered qualified to work on an installation without a supervisor. Security alarm companies wanted to find a reliable wireless security product to install, but the problem was that very few existed.

## Problems with early wireless systems

Because the technology was still in its infancy, wireless products available in the 1970s and much of the 1980s suffered from a variety of problems. The majority are outlined here.

### Insufficient signal strength and range

One problem was that the typical customer of a wireless system owned a house of 20,000 square feet or larger. The normal distance an alarm signal could travel in the 1970s was approximately 100 feet. In addition, the plaster-backed-by-metal-lathe wall construction of a typical 1970s home reduced the effective range of the signal. The technology of the day had little success sending radio signals of sufficient strength and sophistication to provide reliable security.

### Battery life

Early wireless systems were power hogs, and the batteries were not what they are today. The service headaches presented by an alarm system containing 30 or more batteries were many. No one really knew how often to change batteries. And service technicians wondered whether to change all the batteries if they found one dead one.

### Supervision

Early systems weren't supervised, so it was impossible to tell when a battery had run out of power. Alarm companies and customers were not comfortable with the unknowns presented by this problem.

Each month, wireless companies would send service technicians to every customer to check batteries, and they had to build the cost of these checks into the price of the alarm system. And, of course, a battery could check out okay one day and be dead a week later.

### Zones

Simply put, there were no zones in early wireless systems. If a system required 30 fire sensors, they all reported as one. Firefighters wouldn't know whether the fire was in the basement or the attic, or on the east side of the house or the west. They simply received an alarm.

From a fire-reporting standpoint, one zone isn't an enormous problem, inasmuch as the dispatching of fire vehicles to the scene expeditiously is the primary goal. However, knowing exactly where a fire has started can help save time and, in the very earliest stages of a fire especially, can help save lives.

Zone information also assists the service people. If a problem with a sensor develops, which can happen with wired as well as wireless sensors, knowing which sensor to repair is important. Without zone information, troubleshooting a problem is time-consuming even for an experienced technician.

## Radio frequency and false alarms

Wireless systems of the past used only 40-MHz frequency. Systems that operate on 40 MHz have the greatest difficulty penetrating building materials. They are also more prone to false alarms. Taxicab radios, police and fire vehicles, ham radio operators, and high-wattage CB users transmit in the same basic range and therefore tend to affect these lower-range systems. Systems that transmit over the 40-MHz frequency are usually less expensive and less reliable.

## Interference

Because of the frequency used to transmit signals, early wireless security alarm systems were more prone to false alarm and interference. A taxicab, police car, fire truck, or ambulance driving down the street calling their dispatcher could set off alarm systems using the lower frequency. We can remember going into an alarm supply warehouse and receiving calls on hand-held VHF radios. Upon keying the mike to talk back to the office, every wireless alarm system on the rack would false-alarm. Even airplanes flying overhead and ham radio operators could false-alarm these systems. With that kind of problem, it's no wonder alarm dealers steered away from installing wireless systems.



## Transmitter size

Another factor that limited sales of the early wireless alarm systems was the physical size of the transmitter. The thought of gluing a wireless transmitter larger than a cigarette box to a window in a fashionable home was not appealing. In fact, early wireless systems were not truly 100% wireless. You could find wireless door and window sensors; however, other devices needed in the installation had to be wired to a wireless transmitter. Aesthetically, this was worse than the big sensor alone. Now you had two large sensors mounted side by side.

## Features

Few features were available in early wireless systems. For that matter, few features were available in wired systems either—at least not in comparison to what's available today.

## Contention

Another problem with early wireless systems is something called *contention*. In the industry's first attempt to provide supervision of the wireless sensors (the reporting of battery failure and/or sensor malfunctioning), alarm system control panels were programmed to expect to hear from each sensor in the system approximately once an hour. Likewise, the transmitters were programmed to send a radio signal through the air to the control panel every hour.

Contention occurred when two sensors tried to report at the exact same time. When this happened, the control panel couldn't understand either signal. It therefore reported one or both components as having problems, when in fact they both worked.

Contention was a problem for alarm companies because any time a system reported a problem sensor, the alarm company was obligated to send a service representative to the home to fix it. If the problem was caused by contention, the service call is wasted and the alarm company and the customer were both inconvenienced.

## Learning from past experience

The development of advanced wireless security systems in the last 15 years has required that engineers meet several challenges posed by nature of radio technology and the demands of home security. The effective design and function of electronic home security demands that it work in different styles of houses constructed with a variety of building materials laid out in millions of different spatial relations. To provide the fullest possible protection, a system has to communicate among its components, with the system owner, and with a central monitoring station. Yet for optimum protection, the system must only be operable by the system owner, and must be able to communicate its functioning status in a way that makes sense to a nontechnical user.



When the medium of communications is radio-transmitted digital communications, several additional challenges present themselves: the radio transmitter must be powerful enough, have sufficient battery life, signal strength, and range, be of a size that is not obtrusive in the home, and be able to weed out radio transmissions from sources other than its own system.

That's a tall order, but one that wireless manufacturers have successfully filled through years of research and development.

## Breaking the myths

Unfortunately, the issues were not resolved without some false starts. Early systems did have numerous bugs in them. Excessive false alarms, missed signals, dead batteries, interference, and countless other problems gave the new technology a bad reputation, and the challenges faced by designers of wireless security products 15 years ago have led some security dealers to continue to mistrust wireless systems. This mistrust is largely due to myths founded on unexamined biases that favor the tangibility of wired communications over the intangibility of radio frequency communications.

Relying on such myths impedes the progress of the technology and denies customers the often preferable option of a wireless installation. Let's examine each myth and dispel them one at a time.

**Myth 1: Wireless systems are more complex to install and require more programming than wired systems** This is a concern among alarm dealers because if it costs more to install and program a wireless system than a hardwire one, dealers have to pass those costs on to consumers, and that makes them less competitive. However, wireless systems today are extremely easy to install and require no costly labor time for running wires.

And today's most advanced wireless sensors do not require any programming at all. Once sensors are installed, the system is able to recognize the unique sensor identification codes through a simple operation using a control panel touchpad.



**Myth 2: Batteries in wireless transmitters have to be changed constantly** The fact is that systems using long-life lithium batteries typically last 5 to 8 years. A door/window sensor with a battery lasting 20 years became available in 1996. With a battery that lasts this long, battery life becomes a nonissue because the battery is likely to outlast the sensor technology.

**Myth 3: Airplanes flying overhead can cause false alarms in wireless security systems** Sophisticated digital encoding and decoding capabilities of state-of-the-art wireless systems prevent false decoding of signals from other sources, including airplanes.

**Myth 4: A receiver might hear a signal from a transmitter one day and miss it the next.** This myth is based on a time when receivers had only one antenna and a signal could be canceled out if two signals from the same transmitter reached the antenna at the same time. However, today, well-designed systems use redundant, spatially diverse antennas that virtually eliminate the chance of missing signals due to cancellation.

**Myth 5: Wireless systems work only in very small, metal-free buildings** In fact, airports, music theaters, and arenas are among the many applications where wireless burglary, fire, and access control systems provide protection despite long transmission ranges and metal construction. Residential systems are so powerful that range is never an issue.

**Myth 6: Wireless systems aren't reliable** Many hardwire dealers are not familiar with wireless security products, but they hide their lack of understanding by saying wireless systems are unreliable. The best evidence to the contrary would be the millions of homes and businesses equipped with wireless alarm systems that are working perfectly well.

**Myth 7: Wireless systems aren't accurate** Supervised wireless systems provide pinpoint accuracy of detection. When a smoke sensor is activated, for example, the control panel notifies the central station not only that a fire has broken out, but where in the building the fire is located. When subsequent smoke sensors are activated, authorities will know in which direction the smoke is spreading. Supervised wireless systems can also track the movement of intruders as they activate perimeter sensors followed by interior sensors.



**Wireless systems are not supervised** As mentioned, old wireless systems posed problems when batteries died without giving any notification. The dead battery left the door or window unprotected and the end user without a way of knowing whether a sensor was operating properly. Today, supervised systems monitor the condition of batteries and indicate the location of sensors with low battery power long before they fail. Transmitter codes allow users to locate faulty transmitters, PIRs (passive infrared motion detectors), and door/window sensors. Codes can also designate the function of transmitters by programming them for perimeter break-in, interior burglary, fire/smoke, panic, medical, or environmental alarms (water leakage, rate of rise heat sensors).

## In the next chapter

As we have seen, many of the misconceptions about the efficiency and reliability of wireless systems are based on problems that occurred decades ago with old technology. In the next chapter, we'll look at today's wireless systems and how these problems were eradicated.

# 4

# Wireless security: The present

According to the *Security Sales 1996 Fact Book,* 15.4% of burglar alarm systems installed in 1995 were wireless. In 1995, an estimated 333,000 wireless residential security systems were installed in the U.S. Of these, 70%—or 233,000—were installed by professional security dealers.

## Costs of wireless systems

In recent years, the average cost to consumers of residential alarms has dropped from an average of $1,250 in 1993 to $1,100 in 1995. The wireless security industry has shown steady growth for the past five years, and there is no sign of letting up. *Security Sales* estimates that the residential burglar alarm market will climb to $4.2 billion by 1997.

Wireless security is strengthening its hold on the home-security market, with some industry watchers estimating a 50% share by 1998. An increasing number of professional alarm installation companies are switching to wireless security, especially companies who sell a high volume of systems each month. The reason for the shift to wireless is largely economic. Now that wireless alarms are as reliable—even more reliable in many respects—as hardwire systems, companies are finding that the difference in training and labor costs for installers of wireless versus hardwire is significant. Many hardwire companies are owned by electricians who follow the traditions of the trade, including a lengthy apprenticeship for all new installers. Therefore, training hardwire installers is time-consuming, taking upwards of two years for an apprentice installer to become qualified to do a residential installation without supervision.

In contrast to the hardwire apprenticeship system, installers of wireless products can be trained in a matter of weeks. An installer only needs some basic skills with hand tools to install a wireless system.

One study shows significant differences in the cost of installing wireless versus hardwire security systems. The informal study, conducted at a meeting of 14 security dealers, encompasses the sale and installation of 1 control panel, 1 exterior siren, 2 touch-pads, 3 doors, 15 windows, 2 smoke detectors, and 1 PIR. The results are shown in Table 4-1.

**■ Table 4-1 Wireless vs. hardwire cost comparison**

| | Hardwire | Wireless |
|---|---|---|
| **Equipment** | $425 | $1304 |
| **Labor (@ $40/hr)** | $1300 (32½ hours) | $320 (8 hours) |
| **Total** | $1725 | $1624 |

You will note that the cost of wireless equipment is considerably higher than wired equipment in this typical example. The scales tip in the other direction when it comes to labor, however, so the difference in total price is negligible. Note also that the wireless installation took a quarter the time of the hardwire installation, which suggests that the system owner is paying primarily for technology rather than for labor.



## Reliability and range

Some companies actually bad-mouth wireless technology—not because they know how radio frequency transmissions work, but because they *don't* know how it works and don't want to learn about it. Out of ignorance they will say that wireless systems are unreliable or won't work in certain situations. It is in their best interest to convince customers that wireless systems don't work. After all, their business is in hardwire.

The truth is, reliability is no more an issue with high-quality wireless systems than it is for hardwire systems. Although construction materials used in the home—foil-backed wallpaper, mirrors, metal studs, or thick concrete walls—can affect the course of radio wave transmissions, a combination of sensor power and receiver technology ensure successful communications.

Range is not even an issue, as some wireless manufacturers have extended the range of their transmitters to as much as 5000 feet in open air. Thousands of wireless alarm systems installed today have transmitters and receivers separated by thousands of feet. Airports, condominiums, music theaters, arenas, and millions of homes are among the many applications where burglary, fire, and access control systems provide protection despite enormous distances. (See figures 4-1 through 4-4.) Signals sent by wireless transmitters travel to the receiver either in a straight line or by bouncing off surfaces such as metal or concrete.

Keep in mind, however, that not all wireless systems are equal. Inexpensive systems are inexpensive for a reason. They do not all have the same range, the same false-alarm protection, or the same transmitter/receiver technology.

Another important feature to look for is whether the system is fully monitored by a central monitoring station.

■ **4-1** *Many residential wireless security systems are powerful enough to be adapted for use in commercial buildings.*



## Monitored systems

Fully monitored security systems are hooked up via phone lines or long-range radio transmission to a central station. When an alarm is activated, the station dispatches emergency units. Beware of inexpensive systems that claim to be monitored but in reality do not automatically call a central station. Instead, they are set up to call friends or relatives. The friend or relative is then to contact the police or fire department.

Although such a system might be better than nothing, just how much better is not self-evident, and several pitfalls come with the dial-a-neighbor system. If no one is home at the designated phone number,

■ **4-2** *Wireless alarm systems can be tailored to suit any size of application, from a small home . . .*

■ **4-3** *. . . to a large home.*

there will be no response to the alarm. If the designated home has a power failure, there will be no response. Of serious concern for anyone linked by phone to a neighbor is the false-alarm issue. The boy who cried wolf soon lost what he needed most—a quick and trusting response to a call for help. For that reason, and out of common courtesy, no one wants to bother their neighbors unnecessarily.

By the same token, some people will go to great lengths to avoid embarrassment. They are seriously afraid that an unnecessary call to police will make them look foolish or irresponsible, and despite their earnest intentions to respond to your alarm, they ultimately

■ **4-4** *Several control panels are used in condominiums and other multiple-unit structures.*

might not be able to bring themselves to call it in. Or worse, they might put themselves in danger by visually verifying the alarm. Going to the scene of the alarm can be very dangerous. Intruders confront homeowners approximately 10% of the time. It's during those confrontations that assault can occur, and confrontation is what a good security system is designed to avoid.

It's always best to have police investigate an alarm, as they are trained to deal with such situations and to avoid accidental harm to innocent persons. We therefore recommend that the system you buy be monitored by a professional central monitoring station.

Demonstration kits are available for most if not all wireless systems. During the demonstration by your security dealer, ask if the system is monitored at an approved central monitoring station.

## Other issues

Several other issues arise with the purchase of a wireless security system. The most common are detailed in the following paragraphs.

### Battery life

Battery life in wireless alarm systems is getting longer all the time. With the introduction of lithium batteries, the effective life has already been extended to 5 years and longer. Interactive Technologies, Inc. (ITI) has developed a new transmitter that will work for over 20 years before needing replacement (figure 4-5). Upon its introduction, it came with a 10-year guarantee. In addition to long-life batteries, supervised systems provide early warning of an impending battery failure. Therefore, it is no longer necessary to have a technician come to your home every month to check the batteries. Supervised systems check battery life automatically and report battery status well before the battery runs out completely.

■ **4-5** *As the technology advances, battery life becomes less and less of an issue for owners of wireless alarm systems. This long-life sensor battery lasts 20 years.*

## Zones

Fire and burglary systems are mapped in areas of protection, or zones. A map of your security system identifies the type of sensor, its group number (assigned by the manufacturer, this number identifies the sensor's place among electronic circuits in the control panel), its unique sensor ID number, and its location on the premises. With one sensor per zone, it's a simple matter for the system to keep track of where a fire or burglary is taking place in the system.

## Supervision

The newer wireless alarm systems provide full supervision of all sensors reporting to the control panel. The control panel expects to receive a wireless radio report from each sensor at regular intervals every day of the year. With supervised systems, if a sensor is not heard at regular intervals, a report is sent to the alarm monitoring station. At the same time, the system informs the owner of the problem with the sensor. A system with telephone control can audibly report the problem it has found the next time you dial up the security system. For example, you might hear, "Trouble sensor 32, low battery." If you check the status of the system from home, an alphanumeric display screen may provide the message in plain-English text.

The supervisory function of the alarm system will check for the condition of the battery in each of your wireless sensors and whether the transmitter is working properly. It will also check the main battery within the control panel itself, as well as the phone line link to the central monitoring station.

Supervision not only makes wireless systems just as secure as wired systems, it can reduce maintenance and repair costs by providing a way to identify the location of a troubled sensor.

## Radio frequency

Today's wireless systems operate on one of three frequency ranges. Some, usually the cheaper brands, operate in the lower, more busy channels of 40 MHz. Systems operating in the 300- to 420-MHz range are among the most reliable; they require less power to operate, their battery life is longer, their signal is narrower than the higher-frequency systems, and they penetrate building materials better, thereby reaching their target with greater consistency.

Systems operating over the 902- to 928-MHz range, called *spread spectrum*, send redundant signals over several different frequencies, thus ensuring that the message is received by the control panel receiver. Although highly resistant to interference, spread spectrum signals don't penetrate building materials as well as narrow band signals in the 300- to 420-MHz range. Spread spectrum systems also require much more power to operate, which in most cases reduces battery life.

(We'll go into more detail on radio bands in Chapter 6, "The Technology behind the Magic.")

### Interference and false alarms

Two kinds of system technology prevent interference and false alarms. One is the technique of programming each sensor with only one of a possible 16 million identification codes so that no two sensors within range of the control panel are likely to have the same identity. The second is the design of microprocessors with sophisticated decoding capabilities that simply won't allow signals from the wrong sensor to trip an alarm.



With those two technologies working together, a wireless security system is like a lock with over 16 million possible combinations. Radio transmissions can't "unlock" the alarm system receiver and communicate with it because they don't have the right "combination." A wireless security system can't communicate with another wireless device such as a garage-door opener or cordless telephone because the combinations are different.

These advances in technology, along with multibit data transmissions, crystal-controlled narrow band (300- to 420-MHz) transmitters, and learn mode technology (see below), false alarms due to equipment have all but been eliminated. The vast majority of false alarms are caused by system users, suggesting that the industry has taken care of its responsibility to design reliable equipment but has yet to adequately train or educate system users about the costs of false alarms and how they can be prevented.

### Spatial diversity and learn mode technology

The highest-quality wireless systems today use two antennas in the control panel instead of one for added power to receive signals transmitted by a wireless sensor. The use of two antennas for receiving signals is called *spatial diversity,* and it is a feature you should look for in a wireless system.

*Learn mode technology* is another desirable feature in wireless security systems. Essentially, learn mode technology permits a control panel to automatically program each sensor introduced to it while in the program function mode. Once sensors are added into the system, the control panel can hear only those sensors. Sensors can always be added or deleted from the system. Learn mode technology is available in do-it-yourself systems as well as professionally installed ones.

Before learn mode technology, each sensor had to be programmed individually with an electronic programmer. Learn mode technology is automatic and virtually eliminates error caused by mistakes made in programming.

## Size

Advances in electronic circuitry will one day reduce security alarm transmitters to the size of a postage stamp or smaller. One prototype already exists that reduces the sensor size to approximately 2½ inches by 1 inch. Even now, sensors are small enough to fit on doors, walls, and windows without detracting from the room's decor—at least in professionally installed systems. Wireless door sensors can even be recessed in the jamb of the door (wood frames only) and not seen at all.



## Available sensors

Wireless systems use dozens of different detection devices. Smoke detectors, passive infrared (PIR) motion sensors, rate-of-rise heat detectors, glass break detectors, shock sensors, sound sensors, and carbon monoxide detectors are among the most often used devices in wireless systems. See figures 4-6 and 4-7.

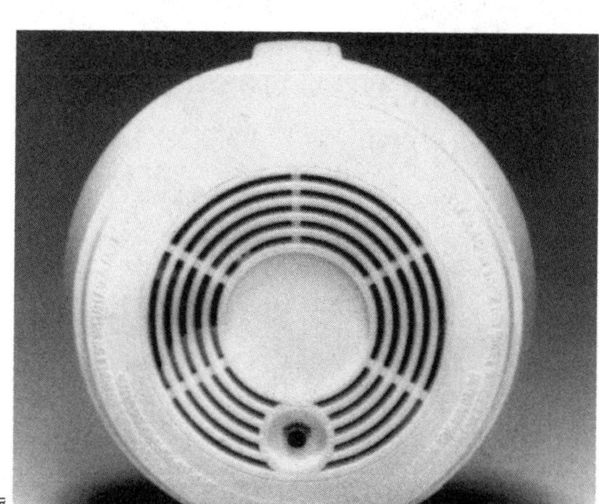
Linear

■ **4-6**
*Wireless smoke sensor.*

■ **4-7** *Wireless motion sensor, or PIR.*



## Consumer Tip: What to Look for in a Good Wireless System

You will find that wireless security systems vary in cost, features, range, and reliability. A system that will give you the security you need with the fewest technical headaches will have features like these:

- ☐ Wireless transmitter ID codes with greater than 1 million settings.
- ☐ Battery life that is greater than 5 years.
- ☐ Crystal controlled frequencies and very narrow band receivers or spread spectrum receivers with high processing gain.
- ☐ Control panels with redundant antennas.
- ☐ Transmitters that are supervised at better than 12-hour intervals.
- ☐ A reputable manufacturer with proven technology and technical support, and a large base of successful installations.

## Features

Wireless technology makes all kinds of features available to consumers. Some of the most widely used features include:

1. *Wireless arming.* With wired systems, consumers were forced to install bulky touch-pad arming stations in various locations of their homes in order to conveniently use the system. Wireless touch pads are portable and can be used to operate the system from locations all over the house. Keychain touchpads that fit in the palm of your hand can be used to arm and disarm your system, bypass the motion detectors, turn lights on and off, open your garage door, or summon police (figure 4-8).

■ **4-8**
*Wireless keychain transmitter.*

2. *Telephone arming and disarming.* With systems that offer telephone control, the need for touch pads is dramatically reduced. All Touch-Tone phones become system touch pads—even ones outside your home. You can arm and disarm your system from your office. By simply calling home, your system will answer and ask for your security code number. You can turn the system off to allow a maid, workman, relative, or neighbor access to the premises. When they are finished, you can call home and turn the system back on.
3. *Energy saver.* Some wireless systems also provide modules that control appliances, lights, and heating/air conditioning. By installing an energy-saver module, you can automate the heating and cooling of your house. You can set the module to adjust temperatures automatically, or call home and command the system to adjust the temperature before your arrival.

   Energy experts say that as much as 30% of energy usage is wasted in heating and cooling an empty house. A feature like the energy saver can save energy without sacrificing comfort.

4. *Medical emergency response.* Medical alert systems are used in nursing homes and private residences to give elderly persons or persons with medical conditions a quick and easy way to call for help during an emergency. Systems include a wireless alarm button that can be worn with a neck strap or clipped to clothing. Some systems have an activity feature that can sense when normal activity stops, indicating a possible medical emergency. A pill-minder feature uses alarm system technology to beep at set intervals to remind system users to take medication.
5. *Two-way voice.* This great feature has helped eliminate a lot of false alarms. Two-way voice modules can be activated by the central monitoring station upon receipt of an alarm. Operators can then listen in to what's going on at the location. If it's a false alarm, the owner simply has to identify him- or herself with an agreed-upon password. Since the operator can hear all of what's going on, the operator can dispatch authorities if there's a lot of commotion or if the person on the premises doesn't report in.
6. *Light control.* Light control can be as simple or sophisticated as you want, ranging from elaborate schemes for timing lights on and off at random intervals, imitating common behaviors in the home, to simplified ways of manually turning several lights off or on from a single control panel button.



## In the next chapter

As you have observed, wireless systems not only work as well as their hardwire counterparts, they also have features beyond those available in hardwire systems. Wireless systems are so powerful that they have been used in commercial as well as residential applications. In Chapter 5, we'll look at some practical applications of wireless security systems.

# 5 Commercial applications for wireless security systems

Wireless security has been put to use in a variety of settings, from ranch houses to airports, from frigid climates to the tropics. Highlighting a few commercial applications demonstrates the versatility of wireless security and how these systems are used to protect people and property.

## Airports



With advances in data transmission technology over the past few years, installations of wireless security systems in airports across the country and around the world demonstrate the power of the technology to perform in the very toughest radio frequency environments. Four features of wireless technology that make it suitable for commercial use also render it trouble-free in residential applications.

1. *Crystal-controlled narrow-band receivers.* These receivers filter out noise from radio sources common to airports, such as voice radio (police, airline, custodial), and weather and navigational radar. The narrow band eliminates interference from stray radio frequency energy, and reliable signal transmission is the result.
2. *Spatial diversity antennas.* (As mentioned, *spatial diversity* means using two antennas instead of one.) With two antennas listening for the transmission, the chances of missing a signal are all but eliminated.
3. *Range.* Distance is often thought to be a critical limitation on wireless technology, yet sophisticated systems have been installed with transmitters thousands of feet away from receivers. Airports are among the many applications where burglar, fire, and access control systems provide protection in spite of enormous distances.

4. *Installation.* Wireless technology translates into fewer installation headaches and expenses for installers and system users. Wireless transmitters allow relatively easy installation and fewer to no costly consultations with engineers concerned with structural integrity and systems maintenance. These considerations are more of a concern on large commercial installations, but they also apply in residential cases where the building is expensively appointed or of historical interest. Because wires do not have to be run, drilling through walls, floors, and ceilings is not required.

A wireless system was attractive to the airport management at Bert Mooney Airport in Butte, Montana, for two reasons: the layout of the airport and a small budget. Forty-four smoke sensors, six fire-pull stations, and two rate-of-rise sensors could be installed in two days, instead of the two weeks required for a comparable hardwire system. The labor for hardwire installation simply wasn't in the airport's budget. Wireless technology also allowed the installation of fire protection in a remote building without running wires under or around runways to the control panel in the main terminal.



Wireless security protects Sydney International Airport from cargo theft. Sensors detect unauthorized movement of cargo in distant warehouses and signal an alarm with wireless transmitters. The particular system in the Sydney airport uses a narrow-band technology that transmits only on a very narrowly defined wavelength, increasing the signal's penetrating power while virtually eliminating false alarms that might by caused by the presence of security equipment.

In addition, wireless shock sensors protected engineers' toolboxes during hangar construction. The toolboxes were simply too large and heavy to move and lock up every day, so portable wireless sensors kept them secure. In the domestic terminal at Sydney International, security guards stationed at metal detectors also have wireless panic buttons for summoning back-up assistance.

A wireless system was installed in the Empress Lounge of Lockheed Air Terminal of Canada, Inc. The lounge had suffered several thefts of valuable electronic equipment and large quantities of liquor. The challenge was to get a radio signal through the four floors of steel-reinforced concrete that separate the lounge from the security command center where the control panel was located. Wireless is flexible enough to work with hardwire components, allowing the option of connecting a hardwire sensor to the control panel via the same route as an existing telecommunications cable traversing all four floors. Placing remote antennas and RF preamp kits to amplify radio signals or repeaters to extend transmissions distances are additional options.

Wireless systems also protect one airline's parts storeroom. Fearing that their spare parts were being stolen and replaced with inferior-quality parts, the airline installed a sophisticated key system and access control; however, neither of these were able to stop the thefts. Then they installed wireless door contacts and PIR motion detectors in the parts storeroom. They also installed sensors to protect the operations center, where controls for lights, heat, and fire detection equipment were monitored; the only way to gain access to protected areas is by getting authorization from the security command center. After the installation, the thefts stopped.

Wireless technology was also installed to protect a crop dusting company's collection of WWII vintage planes. The alternative was to tear up the airplane parking area and run wires inside conduit between the trailer and the hangar—an expensive and time-consuming prospect. A wireless system was the ideal solution.

In another application, a flight school's planes were protected with motion detectors. Equipment in the planes is very vulnerable to theft. With a screwdriver and little knowledge, a thief could get radios and other communications and navigational equipment out of a plane in 30 to 60 seconds. Each plane now contains a wireless motion detector that is activated at the end of the training day. Any disturbance to the plane is now detected by the security system. Because wireless systems can be installed without any modification of the airplane's wiring, they make airplane protection much simpler and easier than hardwire technology.

At Little Rock Airport, sensors were installed to prevent unauthorized personnel from gaining access to runways. The system was designed and installed in response to the Federal Aviation Regulation 107 Security Program, which sets standards for airport security across the country. The program went into effect to combat the safety and security risks created by the size and complexity of most airports. For example, a runway can be difficult to distinguish from a service road at night, and maintenance crews had been known to drive in the landing path of incoming jets. In a different kind of case, a job seeker who had asked for directions to the personnel office learned that it was on the other side of the airport and walked across the runways to get there!

## Historical buildings

Maurice and Christina Lipsey and their two children live in downtown Memphis. The building—formerly the Hart Saddlery Building—was erected 88 years ago and was renovated by Maurice to accommodate Security Watch, Inc., Maurice's home automation and security systems business, and Security Watch Central, Inc., a central monitoring station that monitors some 11,000 security systems in 37 states. Christina's parents, Eva and Jim Moore, also work for Security Watch, and they occupy a second-floor apartment in the building. Maurice, Christina, and their kids live in the 2500-square-foot, third-floor apartment.

Maurice's building is an integral part of Memphis' downtown revitalization project. On land originally owned by Andrew Jackson, it is on the National Register of Historic Places (figure 5-1). When the Lipseys moved in, Maurice had 10,000 square feet of space to protect, but so did the National Register. The building's neighbors have names such as Automatic Slim's Tonga Club, the Memphis Blues Museum, and the Peabody Hotel. In an urban setting where thousands of people traipse across your front yard (in this case, a public sidewalk) every year, it takes more than good fences to make good neighbors. It takes a security system.



The concern for maintaining the look and feel of the buildings' nineteenth-century origins meant that a hardwire security system was out of the question. Maurice's interest in preserving historical integrity of the buildings—and his interest in providing security to his family and his employees—led to the installation of two wireless security systems.

Wireless security offered the best protection for many reasons. A wireless system would cause less damage during installation and would have very little visual impact on the building. "The building has brick walls and the first floor has 25-foot ceilings," Maurice said. "There was really no place to run wires, and I wasn't interested in the expense of renting scaffolding for a hardwire installation. Besides, anywhere we would have run wires, you would have seen wires." Neither he nor the Preservation Society wanted that.

Maurice installed an ITI Commander 2000 for the first-floor lobby and art gallery, where works by local and regional artists are displayed. The lobby is also protected with an access control system and two CCTV cameras. If security for individual works in the gallery is needed, Maurice is able to install up to 17 sensors to de-

■ **5-1** *This building in downtown Memphis is on the National Register of Historic Places. Wireless systems preserve the historic fabric of old buildings.*

tect intrusion or disturbance of art objects. The sensors are set to signal a central monitoring station that is located just upstairs (in most cases, it would be located miles away) from which an operator can notify police that a theft is in progress.

Throughout the 5000 square feet of the second and third floors, Maurice's installer, Rob Hernandez, located wireless smoke detectors, door/window sensors, glass-break sensors, smoke detectors, and motion detectors (PIRs)—all communicating with an ITI Security Pro 4000 control panel. Arming and disarming commands allow free movement throughout the second floor during daytime work hours when the office is running at top speed. The evening configuration is different, as nighttime operators have no reason to occupy areas where motion detectors are turned on to scan the front half of the second floor.

If Maurice is ever uncertain of the arming levels of the second or third floors, a simple call on a Touch-Tone phone—from the bedroom or from across town—will give him a status report of the Security Pro 4000 and all sensors.

"When you do business in the same building where you live, access control is a crucial feature of the security system," Maurice explained. "We have employees in the building 24 hours a day, we have business visitors in our living areas, we have an art gallery on the first floor that attracts the public, and of course, we have friends and family visiting all the time."



To maintain proper security Maurice gives individual access codes to employees, who are trained in the use of the system's touchpads, and the system automatically keeps track of who comes and who goes. His family carries Corby programmable chips that provide access only during specified times of the day, as dictated by individuals' needs. A simple touch of the Corby chip to the chip reader is enough to activate the electronic door strike, sparing 7-year-old Parker and 11-year-old Taylor the common struggle with stubborn keyed locks. The electronic system also keeps unauthorized people out of the stockroom, where thousands of dollars in electronic equipment are in inventory.

Living where you work takes some getting used to. "My office is the living room, the kitchen, the bedroom," Maurice says. Aided by a walkie-talkie, he can be reached anywhere in the building or, for that matter, in the city. And with his laptop computer he can send faxes to system installers from anywhere in the building, which is amply supplied with phone lines.

Although Maurice doesn't always distinguish between his work life and his home life, his security system does. ITI's VuFone is the centerpiece of the home automation system that allows Maurice or Christina to program scripts for coordinating lights, heat, AC, appliances, and security system with family activities (figure 5-2). "Weekday Mornings" is a script that gets the coffee brewing by 6:30, the Classical channel via satellite softly beaming over the living room television and surround-sound speakers, and bedroom lights coaxing the kids out of bed. Motion detectors are turned off, as are the foyer lights.

■ **5-2** *Smart phone interfaces gives system users the advantages of integrated control over security, heating/AC, and lights.*

"Parker's Saturday Morning" script has lights out in his bedroom by 9 on Friday night and cartoons playing by 8 AM on Saturday morning. The elevator door chime, compliments of the Security Pro 4000, is always on, working like an automatic doorbell whenever anyone takes the elevator higher than the second floor. With the script, "Evenings at Home," kitchen lights are out by 10, the TV is programmed to be off and stay off (only the parents know the override code), and bedroom lights are out by 9. Central air or heating are set for comfortable sleeping and timed to anticipate temperature needs in the morning.

## What's in the future?

As more and more businesses go online, the VuFone will give the Lipseys the ability to check bank balances, transfer funds between accounts, pay bills, conduct fast, convenient home shopping, and reach online brokerage services.

But for right now, a digital directory in the VuFone eliminates most of the need for a phone book, and the one-touch dial feature eliminates the chance that Maurice will reach a wrong number because of hasty dialing during a hectic business day.

Christina and Maurice both use the smart phone's Caller ID feature. Maurice knows whether he's getting a business or a personal call. By knowing who is calling, he can screen out unwanted solicitations. If the call is for someone who is not home, he can let the answering machine take the message.

If Maurice weren't in the home systems business, would the Lipseys have the automation and security measures they have come to enjoy? Absolutely. The Lipseys know from experience that they are vulnerable to fire or burglary whether they are in the business or not. Prior to the move downtown, fire struck their home's electrical system and the central station was automatically notified. Firefighters arrived at the house before loss of life or extensive property damage.



They also know how frequently burglary alarms come into the central station every day from among the 10,000 systems they monitor. Accordingly, they have designed a system to provide the most protection with the greatest flexibility. They chose a supervised wireless system because it reports emergencies and its own operating status. Sensors report on their functioning power several times daily, preventing the possibility that a troubled sensor would go unnoticed. During an alarm, intrusion and fire sensors report unique identification codes to the control panel, which, in turn, tells a central station operator the exact location of the fire or break-in (figure 5-3).

According to Carey McNeme, president of Security Watch Central, it's critical for a system to be able to pinpoint the location of an alarm:

> We can track the movement of a fire and let the fire department know, for example, that the first sensor went off in the basement and the second one in the back half of the first floor. In the event of a break-in, it's going to get the police to the scene faster. Let's say we get a report of a window open in the third floor master bedroom. That window might have been left ajar by mistake. We can call and verify what the situation is—either by phone or by two-way voice—and if we don't get a

proper access code from someone on the other end, we dispatch the police. Our information lets police know that they're not dealing with a false alarm and that gets them to the scene faster.

■ **5-3** *Monitored systems dial in to professionally staffed central stations. Operators pass on information about intrusion, fire, and medical emergencies to appropriate authorities.*

McNeme explains that two-way voice activates automatically in the Lipseys' apartment if an alarm is sounded. At the alarm the operator can hear evidence of a burglary, such as slamming drawers or talking between thieves. The operator can then talk directly to intruders, saying, "You have been detected. Provide the proper access code." If there is no response, the operator adds, "Leave the premises immediately. Police have already been dispatched."

Should children, parents, or grandparents ever have a medical emergency, the two-way voice would allow an operator to stay in voice contact until emergency personnel arrive. And at night, when everyone is home for good, the elevator comes up to the second floor and stays there, like a modern drawbridge.

### Business as usual

During business hours, perimeter security and motion detectors are deactivated from phones in the third-floor apartment or the second-floor offices. Maurice disarms the Security Pro 4000 system from the Touch-Tone phone at his bedside, deactivating all third-floor door sensors and motion sensors. Over the phone the system reports in a digitized voice, saying, "Alarm system is off." To prevent unauthorized users from disarming the system, an access code is required for any change in system status that reduces the amount of security in the building. Second-floor security can be armed or disarmed from the second-floor phones as well, providing the flexibility and control the Lipseys desire.

At night, when the pace slows down, Maurice or Christina use the phone to rearm all perimeter sensors. To allow the coming and going of employees, Maurice can bypass specific sensors without having to disarm the entire system. Key holders simply touch their Corby keys to the access control device to activate the door strikes themselves. The Lipsey kids love the Corby keys because they're not like the keys on their friends' rings.



## Museums

The challenge of securing museums is in protecting priceless works of art while meeting all of the expectations of the public and the museum. The public expects an art museum to reinforce feelings of freedom and access rather than of constraint and restriction. Who wants to look at paintings through bullet-proof glass or be viewed by obtrusive surveillance cameras while contemplating the beauty of a seventeenth-century landscape? Bullet-proofing and overt CCTV are for banks and convenience stores, not institutions dedicated to preserving and displaying cultural history. On the other hand, museum curators and insurance companies want art objects protected against theft, vandalism, and unintentional damage. These opposing needs create serious challenges for security.

Most museums have rotating galleries in which works of art come and go throughout the year. Because of the value of some of the exhibits, insurance companies insist on museums providing electronic security for their works of art. If a museum wants to show 40 paintings normally housed in a museum across the country, the insurance company for those paintings would require that they have sensors installed to protect them against theft and vandalism.

If a museum considers a hardwire system, they have the logistical problem of trying to get wires to each painting. To do that they have to assume that the paintings are going to stay the same size. But in fact, because museums have rotating galleries, the paintings move every year and there is no guarantee that a large space won't be occupied by a smaller painting some time in the future. As an alternative to fishing wire that comes out in the middle of a blank wall, the more portable wireless sensors are preferred.

The installation isn't much different from installing a system in a home, except there are more sensors. Museums prefer security systems that can be installed without having to move artwork and without having to close the museum during installation. Moving the art objects is time-consuming and costly in labor, and any movement of the art brings some risk of damage. Having to close the museum for the installation would mean lost revenues, which most museums cannot afford (figure 5-4).

■ **5-4** *Short installation time makes wireless security attractive to museums.*

Wireless systems have additional benefits where climate is a concern. If you're in a climate where it's common to have power outages and temperatures of –30°F, for example, wire strung through exterior walls can be the cause of false alarms.

Some of America's treasures that need security protection aren't in museums. Trade shows and conventions, warehouses, airports and car lots featuring vintage cars and airplanes, computers, costumes, and crafts are potential markets for wireless systems that can be packed up and used at more than one location.

In an especially creative design that helps preserve paintings and conserve energy at the Baltimore Museum, Art Lapole, of Machinery & Equipment Sales, Inc., installed wireless PIR motion detectors that help control lighting levels in galleries. Light has a deteriorating effect on the surface of paintings, so anything that can be done to dim lights when no one is in the gallery spares the paintings and reduces the museum's energy costs. Art installed wireless motion detectors in galleries to detect the presence or absence of visitors. When a visitor leaves a gallery, the PIR signals the electronic dimmers to lower the lighting level. Dimmers slowly raise the light level when the gallery is reoccupied.



## Schools

School divisions that invest millions in buildings, computer, AV and office equipment, books, and furniture every year have a need for security that will keep these publicly funded institutions safe from burglary and vandalism. Schools are as vulnerable to theft and vandalism as any other institution, perhaps more so when the risk-taking behavior of some adolescent students is taken into account. Schools in remote sites have the additional challenge of procuring security systems that can be installed, serviced, and monitored without incurring prohibitive costs.

These were among the several concerns of the Fort Vermilion School Division in northern Alberta. The need was to establish perimeter and interior protection against theft or damage to libraries, computer centers, AV rooms (equipped with projectors, TVs, and VCRs), science labs, and offices before any damage or loss occurred.

When Fort Vermilion began their search for a system and a dealer who could provide the security they needed, one very specific qualification had to be met: they didn't want to pay a dealer to install the 17 systems they planned to buy. They also wanted a system that was versatile enough to work in buildings that had some existing security systems in place or whose construction posed some challenges to the installing dealer.

Buildings offered installers no ceiling access and no basement access for running wires. In addition, the large square footage of the schools would have meant high labor bills for wiring, and the wire would have had to be exposed. Wireless was required for two other related reasons: 1. the design of some buildings meant having to run wires and conduit down hallways; and 2. the students use these same hallways. In addition to the unsightliness of exposed wires or pipe, a hardwire system would have been vulnerable to vandals and pranksters.

Schools require a great deal of versatility. Room uses can change yearly, and when a classroom is converted into a media center where 10 to 50 computers will be stored and used, the security needs of that room change too. With wireless all it takes is the removal and replacement of a motion detector from one room to another. The same flexibility is not always available with hardwire, as a comparable move would require paying labor costs for drilling and running wire from the old location to the new. However, a dealer—or in the case of Fort Vermilion, a staff engineer—could pluck a wireless PIR off the wall in the old computer room and reinstall it in the new one in a matter of minutes. Wireless equipment is easy enough to manage that building engineers could also install add-ons should the need arise.



The school division of Fort Vermilion was already sold on wireless when Horizon got the go-ahead. The WatchGard systems installed could be hardwire, wireless, or a combination of both. The system's technology allows it to be controlled from traditional hardwire touchpads, wireless touchpads, or most Touch-Tone telephones on or off the premises. Mobile classrooms were protected with wireless PIRs and door sensors, and the control panels were located in the main-building mechanical rooms. Wireless PIRs would protect the mobile units and beam alarm signals to a control panel several hundred feet away.

Schools are well suited to wireless because the technology keeps costs in check. Horizon sold all 17 systems at once, which was good for the dealer, but the division's costs were lowered for labor because they already had the electrician on the payroll. He could install the systems as time permitted.

The WatchGard system gave the school division telephone control over the arming of systems—a feature whose value increases with the distances between schools. School administrators liked being able to arm and disarm the system and check on the status of batteries and sensors over the phone.

## Farms

When David Wolff and David Albin of My Alarm Company, Ltd., in Centreville, Virginia, were contacted by Euphemia and Rufus Hutcheson, all they knew was what their telemarketing list told them: the Hutchesons were over 55 years of age and made over $55,000. They had no way of knowing about the Hutchesons' interest in horses. Interest is perhaps an understatement—passion is more like it. Like the clients who board their horses and keep their livery at Misty Brae Farm, the Hutchesons have put uncountable hours and thousands of dollars into their love for horses. It shows in the breeding of the horses, in the care of the stables, and in the huge inventory of saddles, bridles, blankets, cinches, and boots to be found in the tack room.



When the two technicians arrived at Misty Brae, the Hutchesons told them about comments that two strangers had made during a visit to their riding arena. Incredibly, the two strangers made inquiries about the farm's security and started asking questions about cattle. Misty Brae is not in cattle country, and the questions seemed like idle chit-chat designed to buy time while these two visitors sized up the layout of their property. The incident was strange enough to raise the Hutchesons' security concerns to a new level. In their 24 years as horse owners, they had never thought about security at all. Now that these two men had paid their odd visit, security was a top priority.

The Hutchesons wanted a system that would notify them of unauthorized entry into the 200-by-200-foot riding arena where stables and riding equipment are located. The walls and roof of the arena are made entirely of corrugated metal. Horse owners and trainers would need access to the horses day and night, and the Hutchesons wanted to be able to control the system from the house, which is 300 feet away.

When the Hutchesons mentioned that they also wanted security for a nearby barn, which was built in the 1880s and was originally pegged together without metal nails, David Wolff got the impression that they were proud of the history of the structure. Accordingly, he drove home the point that wireless transmitters could not only cover the range between the barn and the metal riding arena and the house, but that no trenches would have to be dug and no conduit would have to be run in, around, or through any of their farm's buildings.

The Hutchesons were also concerned about damage to the system from any of the various animals that inhabit a farm. Rats are often a part of life on a farm, for example, and they can cause serious damage to wires by chewing through them. Rats can't chew through air, though, and that also made wireless very attractive to the Hutchesons.

To satisfy Misty Brae's need for security and fire protection, Wolff installed wireless rate-of-rise sensors over the hay bale storage area. There was too much dust to use smoke sensors, and no matter how well protected a smoke sensor might be, spiders can find a way to get in and spin webs that set off false alarms. He also installed wireless temperature sensors above the office where power lines enter the riding arena. Door switches did the job on the doors to the riding arena and the barn. PIRs were out of the question, however, because whole flocks of sparrows had free reign in the riding arena. A wireless touchpad in the house gave the Hutchesons the control they wanted over sensors located in all of their outbuildings.



The system added value to the boarding services provided at Misty Brae, as the monitored system gives tenants the ability to summon a doctor or a veterinarian with the push of a button, day or night, in the event of fire, serious illness, or a riding accident.

Placing sensors in the riding-arena doors—big enough to drive a crane through, according to Wolff—proved to be the biggest challenge to the installation. David had to place the magnets in the wood flush with the surface to prevent them from being sheared off at the door frame.

State-of-the-art wireless systems do their jobs despite building materials and environmental factors common to farms, namely metal roofs, long transmission distances, exposure to the elements, and the presence of creatures great and small.

## In the next chapter

In this chapter, we saw how wireless systems are used in a variety of applications to ensure reliable and efficient security. In the next chapter, we'll look more closely at the technology and define a few terms.

# 6

# The technology behind the magic

Wireless security systems employ the same principles of communication at work in your television and radio. The medium that makes such forms of communication possible is the electromagnetic wave and the devices that produce them, send them, receive them, and interpret them. When a wireless sensor in an armed security system is activated, a complex communications process is set into motion. Each security system sensor has two basic functions. The first is to detect a change of state; the second is to send a message about that change of state to the control panel via radio-frequency energy.



## How wireless systems work

*Frequency* is the number of times per second that a radio wave completes a cycle. *Wavelength* is the distance between the beginning and ending of the wave. *Amplitude* describes the amount of energy the wave produces, shown graphically as the height of the wave above the baseline on an oscilloscope. The radio spectrum is from about 30 kHz (kilohertz, or thousands of cycles per second) to 300 gHz (gigahertz, or billions of cycles per second).

Because so many radio-frequency messages can be generated by anyone with a transmitter, the Federal Communications Commission (FCC) has assigned certain functions for specific frequency bands on the radio spectrum. For example, navigational and aeronautical communications use lower bands of 30 to 300 kHz, whereas long-range communications from satellites and land microwave transmitters operate in the 3 to 300 gHz range.

A wireless security system is made up of several radio transmitters that are designed to send very specific messages over radio waves to the receiver, which is in the system control panel. Security-system transmitters encode information by modulating an electromagnetic wave, which radiates out in all directions from the transmitter's antenna. A comparison is often made between radio waves and the waves that radiate out when a pebble is dropped in a pool of water. The wave continues radiating until it hits an obstacle or runs out of energy. The waves transmitted and received by a wireless alarm system are much more sophisticated than the water example suggests, however. They have a much longer range, and they vary greatly in how much power they consume. The pebble-in-the-water analogy also doesn't show how a unique message is impressed upon the wave.

Battery power applied to the transmitter antenna in each sensor sends electromagnetic waves, which propagate through space and reach the receiver antenna. In this respect, the radio wave is a communications medium that is also the message. The shape of the wave is modulated in very precise ways by the transmitter, constituting a unique message. The radio receiver is able to accept the modulated signal and interpret the meaning of the waveform modulation. When the waves reach the receiver's antenna, the receiver interprets or decodes the received information, translates it, and passes it on via phone lines to the central monitoring station in a form that is useful for the central station operator.



The transmitter will send information about which sensor is transmitting, whether it is an intrusion, fire, medical, or environmental sensor, and what its location is. That same information is passed on to the monitoring station. The operator receives notification of an alarm on a computer monitor, which displays the type of alarm and, depending on the type of security system, which door or window has been broken into or where in the house a fire has started.

Receivers programmed only to listen to these unique signals ensure reliable system communications. They also prevent false alarms by filtering out signals from the wrong sensor (say, from a neighbor's house). When you have a wireless system professionally installed, the installer will spend time adding sensor numbers to the system. The system will thereby be programmed to identify only those sensor numbers so it can't activate a false alarm caused by a non-system radio signal. Old garage door systems and early wireless security systems didn't have such sophisticated communication safeguards and could therefore communicate with—and activate—each other.

Wireless alarm systems do not all employ the same communications technology. Wireless systems can be grouped into two main categories based on their transmission frequency: narrow band or spread spectrum. *Narrow band systems* send the signal via a single frequency. By transmitting at only 319 MHz, for example, a narrow band system sends its message so that only a receiver tuned to the same very narrow bandwidth will be able to receive the message. An analogy would be a radio that is designed only to pick up one station—say the Oldies station—and would not be able to hear stations that broadcast classical, rock, jazz, or talk radio on different frequencies.

A *spread spectrum system* works by transmitting across a wider band of choices on the radio spectrum. A spread spectrum system will transmit the same data simultaneously on over 100 different frequencies around the 900-MHz range. Because of the redundancy of the message transmission, the signal has an excellent reception. Even if part of the signal is lost due to interference from cellular phone or pager transmissions on the same wavelength, the message will still get through to the receiver.

Although the technology has been available since the 1940s, spread spectrum systems didn't reach security dealers until the last few years. Spread spectrum is used with much of the most sophisticated military radar and communications equipment around today. One of the key features of a properly designed spread spectrum system is its resistance to jamming.



Short-range wireless security systems—either narrow band or spread spectrum—operate at signaling frequencies in the ranges of 40 to 50 MHz, 300 to 450 MHz, and 900 to 930 MHz. Equipment designers select frequencies to optimize transmitter cost, size, battery life, and performance. Narrow band usually means smaller transmitters and the consumption of less power than spread spectrum. Spread spectrum's advantages include higher data rates and the ability to transmit at higher power levels. The technology excels in inventory-control applications where many terminals are required or where messages are long.

Narrow band transmitter/receivers can operate at 300 MHz, which is a very quiet and little-used band that is vigorously protected by the government. 300 MHz also has especially good building-penetration capabilities. Spread spectrum systems have a longer range than narrow band systems, but narrow band range is so long that it can achieve any range necessary in a residential application.

## Narrow band versus spread spectrum

"Narrow band" and "spread spectrum" refer to the way a message is placed on a radio signal. A narrow band system uses a small part of the spectrum for messages, while a spread spectrum system spreads the message over a wide frequency range.

***Advantages of narrow band systems:***

***Smaller transmitters***

***Fewer components***

***Longer battery life***

***More features***

***Less interference from RF sources***

***Advantages of spread spectrum systems:***

***High resistance to jamming***

***Higher data rates***

***Excellent penetration through building materials***



Each sensor in a system emits a radio wave that modulates at a certain frequency and in a very specific way. The transmitter is designed to alter the wave in specific ways for each message, and the receiver is designed to interpret the alterations in the wave to mean specific things. Depending on the message from the various circuits of the detection device, the sensor's transmitter formulates a message that is carried by the transmitter's radio wave. To actually create a message, the transmitter takes information from the detection device in the sensor and alters the shape of the carrier wave. The message that the carrier wave sends to the receiver may include the status of the sensor (e.g., whether the door it is monitoring is opened or closed), the status of the sensor battery, and an identification code for the sensor, which will identify its location on the premises.

To avoid the risk of missing a signal from a transmitter, some systems employ spatially diverse antennas to virtually eliminate the chances of missing signals. With one antenna, there's a chance that two or more waves from the same transmitter will reach the antenna at the same time. That is not in itself a problem, but if they both reach the antenna at the same time and their waves are in opposite positions to each other, the message can be canceled out. But by polling between two antennas instead of one, the control panel can always read one or the other of the messages that reach the antennas.

Wireless transmitters can be separated from receivers by thousands of feet because they have sufficient power supplies and internal controls to sustain the message. Supervised wireless systems also provide pinpoint accuracy of detection. When a smoke sensor is activated, the control panel notifies the central station not only that a fire has broken out, but where in the building the fire is located. When subsequent smoke sensors are activated, authorities will know in which direction the smoke is spreading. Supervised wireless systems can also track movements of intruders as they activate perimeter sensors followed by interior sensors in their search for valuables.

Old wireless systems posed problems when batteries died without giving any notification. The dead battery left the door or window unprotected and the end user without a way of knowing whether a sensor was operating properly. Supervised systems monitor the condition of batteries and indicate the location of sensors with low battery power long before they fail. Transmitter codes allow users to locate transmitters, PIRs, and door/window sensors should they ever be in need of service. Codes can also designate the function of transmitters by programming them for perimeter break-in, interior burglary, fire/smoke, panic, medical, or environmental alarms (e.g., water leakage, rate of rise heat sensors, etc.).

Engineers at wireless manufacturers are noted for innovation in meeting the demands of commercial, residential, and government security. They have designed transmitters to protect paintings, to arm and disarm security systems, and to signal changes in air pressure or the presence of body heat.

## The language of wireless

Like other areas of specialization, wireless technology has its own vocabulary. Understanding how wireless works requires learning the language of RF energy. Following are a few terms you should know. Although these terms, and many more, are included in the Glossary, we wanted to highlight these particular terms so that you will be familiar with them before continuing.

*carrier* An electrical oscillation that is capable of propagating through space, from the transmitter to the receiver, without wires.

*crystal controlled oscillator* An oscillator whose frequency is determined by a quartz crystal. Used in transmitters, these types of oscillators have the highest frequency, accuracy, and purity.

*diversity* The use of redundant signal paths. For example, the SX-V control panel by Interactive Technologies employs two antennas instead of one. Although the odds of missing a single path signal are generally very low, it can occur. With two antennas, the odds of missing a signal traveling two different paths at the same time are exceptionally low.

*free-air range* The unobstructed communications range of a transmitter-receiver system.

*interference* Anything that reduces the communications range of a transmitter-receiver system.

*narrow band* Transmitters that send with a very narrow bandwidth radio signal and have 10 times less chance of being interfered with than spread spectrum systems. Also a term that has special meaning in the alarm industry: it means any radio that is not spread spectrum. In a narrow band system, the smaller the receiver band width, the less chance the signal will be interfered with.



*receiver* The RF analog and logic components in the alarm panel. They process the signals received by the antenna and regenerate the digital message of the transmitter.

*spread spectrum* Transmitters that communicate across a wide band of choices on the radio spectrum. Spread spectrum systems transmit the same data on over 100 different frequencies at the same time.

*supervision* The use of a special signal sent automatically from the transmitter to the receiver at regular intervals to inform the receiver that the transmitter is functioning properly.

*transmitter* A device consisting of a radio frequency oscillator that generates the carrier frequency, a modulation method to impose the messages onto the carrier frequency, and an antenna that radiates the signal.

*wavelength* The distance that the radio wave travels in one cycle of the transmitter's frequency. A 300-MHz transmitter has a wavelength that is optimum for penetrating building materials.

## In the next chapter

Now that you have the basic technology under your belt, let's move on to a more detailed description of each component in a wireless security system. Chapter 7 focuses on burglar alarm systems.

# 7 Components of a wireless burglar alarm system

No discussion of wireless security is complete without an understanding of the components that make up an alarm system, how they work, and in what circumstances they are used. Wireless security systems are designed to detect conditions that indicate intrusion, fire, medical emergency, or environmental hazards and report those conditions electronically—without the use of wires between sensors and the system control panel. Sensors detect opened doors or windows, broken glass, smoke, or frozen pipes, and communicate those conditions via battery-powered transmitters that send radio signals to the control panel. Designing and installing a wireless security system is a matter of assessing security needs, selecting the equipment to meet those needs, and designing a system that suits the equipment, the end users, and the space to be protected.



## Control panel

Depending on the type of system, the control panel will either be hidden away in a closet or utility room, or it will be placed in a convenient location where it can be used to operate the system (figures 7-1 and 7-2). Some control panels have a built-in touchpad for turning the system on and off, for conducting system status checks, and for sending panic alarms. These control panels are mounted for easy access in a living area of the home. System control panels that are not designed for user interface are hidden away, in the same manner as an electrical circuit panel.

The function of the control panel is to supply the electronic intelligence to the system. The control panel derives its power from a plug-in transformer, which converts your normal house current of 110 volts to either 6 or 12 volts, depending on the type of system.

■ **7-1** *Large control panels are installed out of sight. They are connected to a power source, the phone line, and to optional hardwire devices.*

■ **7-2** *Control panels with built-in touchpads are mounted on a convenient wall.*

The best control panels monitor their own power. As long as it sees 6 or 12 volts of electric power coming from the plug-in transformer, the panel continues to use that power. If, for some reason, normal house electricity is lost, the control panel automatically switches over to its internal back-up battery power, which continues to operate the system for several hours or until electricity is restored to the home. Better systems use rechargeable back-up batteries. When they are used to run the system and normal house current is restored, the control panel recharges the back-up battery automatically so it will be ready when the next power outage occurs.

The control panel communicates with a central monitoring station via telephone lines (most commonly) or via long-range radio transmission (a relatively new technology, but one that is gaining ground quickly). From the central monitoring station, an operator dispatches the proper authorities to the scene of the alarm.

## Sensors and transmitters

Sensors consist of devices for detecting changes in environmental conditions and a transmitter for communicating those changes to the control panel. The electronic design of wireless communication employs radio waves as carriers of messages. Those digital messages are generated by the battery-powered transmitters and are "heard" by the control panel—the system's receiver.



The most sophisticated system transmitters generate signals that carry unique codes, which indicate the working status and the location of sensors on the premises. Codes also prevent false alarms caused by signals transmitted from non-security system sources. Since the receiver can only hear certain codes, false alarms cannot be caused by radio signals from sources such as TV remote-control devices, garage door openers, or airplanes passing overhead. Sensor identification codes are programmed into the sensors by technicians or by the manufacturer.

### Supervised sensors

In supervised systems, the signal indicates to the control panel and to the central station which sensor has been activated. A supervised system would tell police, for example, that an intruder entered the upstairs bathroom window, then went into the master bedroom and opened a protected cabinet. In a multiple-unit application, such as a farm or ranch with several outbuildings, an alarm would indicate in which building an alarm has been sounded. In one school application we know of a vandal was found hiding in a portable classroom, unaware that the alarm had pinpointed his location. Supervised sensors also automatically perform communication testing and battery monitoring so that the user is notified of a low battery or a noncommunicating sensor.

Oscillators in the sensor transmitters establish the frequency at which the radio signal is sent. One of the cost variables in a wireless security system is based on the type of oscillator employed in system sensors. Systems whose signal oscillation is controlled by a quartz crystal oscillator ensures the accuracy and stability of the signal. Crystal-controlled systems are more expensive but are worth the money because they are more accurate.

## Door/window sensors

Door/window sensors detect the opening and closing of doors and windows. They can also be used on drawers, display cases, and cabinets. A magnet and reed switch mechanism detects openings and closings (figure 7-3). The role of the magnet is to keep the reeds in the sensor away from each other. When the door is opened far enough for the magnet to stop having an effect on the metal reeds, the reeds make contact with each other and signal an alarm. Better surface sensors also include a tamper switch. The tamper switch is designed to detect if anyone attempts to detach the sensor from the door or window in order to circumvent the alarm system. Tamper switch protection is usually on at all times, so even if the alarm system is off, you would be informed if the sensor is being tampered with.



Some door/window sensors have connectors that permit the attachment of other wired sensors. This is highly desirable and cost-effective when you have groupings of windows that can be protected with one wireless transmitter and a couple of wired sensors connected to it. A bay window is a perfect application for this type of sensor. If the bay contains four windows that open, one transmitter plus three hardwire sensors could be used, saving you the alarm panel memory for other areas of the house.

## Recessed door sensors

Recessed door sensors are slightly more expensive than surface-mount sensors but are often worth the extra money when it comes to aesthetics. A recessed sensor can be installed in the jamb of the door and is therefore out of sight when the door is closed (figure 7-4). Installation of this type of sensor is more difficult than a sur-

■ **7-3** *Wireless door/window sensors have a reed switch mechanism and a built-in transmitter.*

face-mounted sensor and requires someone to install it who is experienced with the product. A long hole is drilled into the door frame, and the sensor transmitter is inserted into the hole. A shallower hole is drilled into the top of the door and the magnet is placed there.

ITI

■ **7-4** *Recessed door/window sensors almost completely disappear.*

## Types of sensor technology



Many types of sensors are available to suit the application. Typical systems use a number of different kinds of sensors, depending on the area of the building to be monitored.

☐ Passive infrared (PIR) motion sensors work by detecting body heat (figure 7-5). PIRs can be adjusted to work in broad or narrow spaces. They can be aimed and masked to prevent false alarms caused by pets, heat vents, and other common causes.

Linear

■ **7-5** *Wireless passive infrared motion detectors provde interior security protection against intrusion*

- ☐ Sound sensors "hear" specific frequencies, such as those made by breaking glass or splintering wood when an intruder uses force to gain entry (figure 7-6). Placed on a wall or ceiling, sound sensors are economical in rooms that have a lot of glass.

■ **7-6** *Sound sensors respond to the sound of breaking glass and are especially useful in rooms with several windows. They are usually installed on a ceiling.*

- ☐ Shock sensors mount on window frames and work as door/window sensors. If an intruder should break the window instead of prying the window frame open, however, the shock sensor detects the shattering of glass as well. See figure 7-7.
- ☐ Smoke sensors detect smoke and then sound an alarm. In supervised systems, the transmitter in the sensor sends a fire alarm to the control panel and indicates the location of the sensor that detected the smoke.
- ☐ Rate-of-rise sensors detect sharp increases in temperature. They are used where smoke or high temperatures might occur naturally, such as in kitchens, furnace rooms, garages, or attics.
- ☐ Although not technically a sensor, wireless panic buttons allow the user to summon help without having to get to a phone (figure 7-8). The wireless transmitter signals the need for help, and the control panel contacts the central station.

■ **7-7** *At a manufacturer's training session, students test the sensitivity of a mounted shock sensor.*

☐ Freeze sensors detect furnace failure before damage can occur (figure 7-9). They have a built-in thermostat that activates an alarm when the temperature drops below a certain level, notifying you of the need for furnace repair before pipes have time to freeze.

■ **7-8** *Wireless panic buttons can be used to call for help when the user is hundreds of feet away from the system control panel.*

## Touchpads for wireless systems

Wireless touchpads allow the user to control the system without having to go to the control panel (figure 7-10). Most operations can be done with wireless touchpads, which are located in convenient places throughout the home.

Hardwire touchpads may have a window with easy-to-understand readouts of system commands or functions (figure 7-11). In some systems a speaker can be plugged into the touchpad for voice confirmation of system functions. During an alarm, the voice announces

■ **7-9** *Freeze sensors have a built-in thermostat to detect when the ambient temperature drops below a certain point.*

■ **7-10** *Wireless touchpad.*

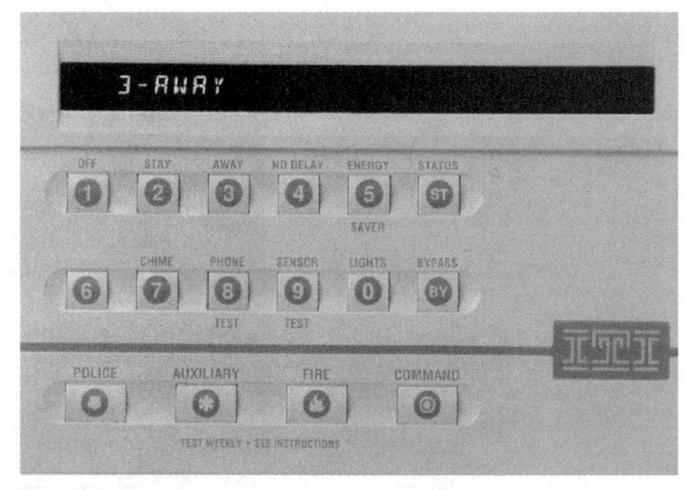


■ **7-11** *Hardwire alphanumeric touchpad.*

whether a fire or burglary sensor has been activated. A fixed-text display shows the arming level and system status so the user can have important information at a glance. (Note that hardwired touchpads may be part of an otherwise wireless system.)

Touchpads can also integrate several communication components into a single unit for easy mounting and user access. In addition to providing voice feedback, a talking touchpad equipped with a two-way voice speaker and microphone can provide alarm verification. Some touchpads also include panic buttons for police, auxiliary, and fire response; a command button for simplified operations; test buttons; heating and lighting controls; and status, bypass, and chime functions.

## Central monitoring receivers

A central station is equipped with receivers that monitor security systems via telephone lines. A receiver such as ITI's CS-4000 is interactive, which allows two-way communication between the control panel and the receiver (figure 7-12). An interactive receiver allows the operator to use keyboard commands to program information into a control panel's memory. For example, commands may instruct a control panel to dial the central station and give its account number along with alarms, trouble, or test reports.

With information gathered via the interactive receiver, a technician can learn the nature and location of a technical problem before arriving at the home where the system is installed. Some

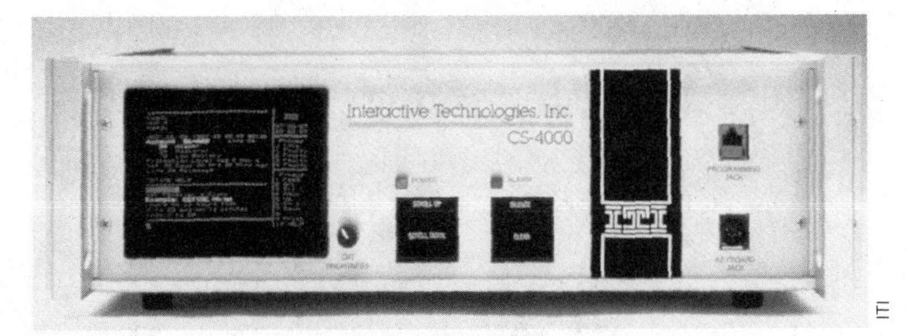


ITI

■ **7-12** *CS-4000 central station receiver.*

simple problems can even be fixed via the interactive receiver, eliminating the need for a house call. The receiver also allows the operator to make changes in the control panel's memory so it can be customized to specific requirements.

## Lamp modules

As mentioned in Chapter 2, staggering the lighting in your home when you're away is much more effective at discouraging intruders than setting everything to go on and off at the same time. X-10 lamp modules (made by X-10 USA) allow the security system to turn specified lights on and off (figure 7-13). Each module has its own address and can be controlled to turn on or off by receiving commands over existing electrical wiring. X-10 modules can be controlled by many security system control panels.

■ **7-13** *X-10 lamp module.*

## Sirens

Wireless sirens used to frighten intruders and notify others of an alarm can be plugged into a wall outlet. These can be purchased off the shelf or provided by your security technician as part of the system.

# Bringing it all together

Now that you have a basic understanding of each component in a wireless security system, let's look at how the components work together. As an example, we'll use an ITI Commander 2000 control panel, a moderately priced system that is commonly available around the country.

To activate or deactivate the system, the user sets the arming level of the system by first entering a private code. The code allows access to the system and prevents unauthorized persons or very young children from operating it. The access code is used any time the user wishes to reduce the amount of security on the premises.

Indicator lights notify the user of system status (figure 7-14). In other words, the system communicates with the user to reassure the user that the intended arming level or other settings have been activated by the system.

■ **7-14** *Control panel with built-in touchpad.*

Numbered buttons on the control panel arm the system for different circumstances. In the STAY mode, for example, the alarm system can be on, but people can move about inside the protected area without setting off alarms. The perimeter doors and windows are armed, but interior sensors that would be activated by movement or heat do not respond (that is, they are bypassed). When in the AWAY mode—meaning persons normally in the house are away—all interior and exterior sensors are on.

NO DELAY refers to the readiness of doors to go into alarm immediately if opened. When NO DELAY is turned off, doors will allow the user a delay time during entry and exit for arming or disarming the system before an alarm sounds.

Panic buttons (the double buttons on the right of the panel) allow quick and easy activation of an alarm. Buttons specify police, fire, or auxiliary (medical) alarms. No matter what arming level is selected, the panic buttons notify the central monitoring station of the need for help when pressed. Panic buttons cannot be turned off or bypassed.

Selecting the chime mode on the panel activates a chime or beep every time an armed door is opened. No alarm will sound, but users know when someone enters or leaves the premises.



Using the COMMAND button on the panel is a quick way of increasing the arming level of the system. Instead of entering a four-digit code and then pressing the button for the desired arming level, the user can press command + 3, for example, and the arming level of the system will be increased to the AWAY mode (all perimeter and interior sensors are on). As a security measure, the COMMAND button cannot be used to turn the system off or to reduce the level of protection.

The BYPASS button allows selected sensors to be bypassed so that an individual window could be opened on a hot night without setting off an alarm.

The STATUS button allows users to check on the operation of the system. If a sensor has a low battery or if a sensor cover is off or a window is open, the system will inform the user of the situation. An optional LIGHTS feature provides control of any lights that are on light-control modules.

Other buttons allow the testing of phone lines or of sensors. Testing phone lines is important because the security system is connected to the central monitoring station via the phone. Sensor testing allows checking on the operation of sensors without the risk of accidentally sending an alarm to the monitoring station.

In addition to receiving signals from sensors, interactive systems such as the Commander 2000 can also be programmed using two-way communication between the control panel and the central station receiver via phone lines. Only interactive systems allow programming and troubleshooting functions to be done remotely. The advantages of interactive technology include allowing the user to set access codes, add or delete sensor numbers, and perform several other functions without a service call from a technician.

## In the next chapter

In this chapter, we detailed each component of a home security system, from the control panel to lamp modules. In Chapter 8, we'll look at the components in a typical fire detection system.

# 8 Components of a wireless fire detection system

Because the dangers of fire are so great and the consequences of fire so devastating, all reasonable precautions to prevent loss from fire should be taken. In the era of electronic detection and almost instantaneous central station reporting, is it reasonable for anyone not to have a monitored fire alarm system? Sure it is, because there's a financial side to the story. Monitoring costs from $18 to $25 per month and, depending on the company you're dealing with, will require a multiyear contract—usually between 1 and 5 years. Some companies will allow you to go on a month-by-month basis, which is true of only some of the do-it-yourself companies that have made arrangements with monitoring companies.

The truth is, many people don't like the idea of having a monitored system. Whatever the reason—they don't trust monitoring stations, they don't like to pay a monthly monitoring fee, or they feel like Big Brother is watching them—they're not convinced of the value of having a monitored fire system.

But the difference between a monitored fire system and an unmonitored one is pretty clear. Although both systems can be effective in warning inhabitants in time to get safely out of the house, an unmonitored system will not call the fire department and report a fire within seconds of detecting smoke. Seconds are crucial when it comes to fire, since poisonous gases can overwhelm inhabitants within three minutes of the fire starting.

Monitoring also helps prevent false alarms and the "cry wolf" problem that arises when your exterior siren goes off so many times your neighbors just ignore it. Monitoring stations will verify fire alarms whenever possible, and you don't have to worry that the only conscientious neighbor will be on vacation the day your house catches fire and your exterior siren goes off. Monitoring ensures that your house is being watched day and night whether you are home or not.

## Basic system

All basic fire alarm systems include two components: a control panel, which sets the system, and a line carrier power transformer, which uses household current to power the panel and wireless interior sirens. In addition, a household fire warning system includes smoke and heat sensors and a fire pull station. Rate-of-rise heat sensors, monitored smoke detectors, and fire pull stations provide optimum protection against loss due to fire by providing early warning to inhabitants and almost instantaneous communication with the fire department.

### Heat sensors

Rate-of-rise heat sensors are designed for use in areas where non-threatening smoke might normally exist, such as near a fireplace or in a kitchen, and where high heat normally exists, such as in an attic. The heat sensor will initiate an alarm when a fixed temperature has been detected or when the rate of rising temperature exceeds a predetermined limit.

Heat sensors, like smoke sensors, are referred to as 24-hour sensors because they cannot be bypassed. They are battery powered with any of a variety of batteries. Lithium batteries are the longest lived, with manufacturers specifying five to eight years of life under typical conditions.



### Smoke sensors

Smoke sensors can be either the ionization type or the photoelectric type. *Ionization smoke sensors* are designed to detect particles of combustion. An electrical current is conducted inside the sensor, and if smoke interrupts the current, an alarm is activated. Ionization detectors provide a faster response to open-flame fires.

*Photoelectric smoke sensors*, on the other hand, generate a beam of light. If that beam is broken, an alarm is sounded. These sensors are particularly effective in detecting "cool smoke," produced when a mattress or sofa smolders from a cigarette. This kind of smoke contains carbon monoxide and can kill sleeping occupants before setting off an ionization detector.

Some smoke sensors feature symmetrical smoke detection chambers for detecting smoke equally well from any direction. Some might also have a self-contained alarm horn, a low battery/sensor failure annunciation, and an indicator light that flashes periodically to indicate normal operation.

### Fire pull stations

Fire-pull stations are the fire alarm system equivalent of a police panic button. They are wireless transmitters in a box that look like the stations you probably remember seeing in school hallways. When the user pulls the lever, the signal is identified at the control panel, and in turn, at the central monitoring station as a fire alarm. Frequently located in kitchens, where fires are common, fire-pull stations are a convenient way for the user to set off an alarm without having to go to the control panel.

## Benefits of a wireless fire alarm system

Wireless fire alarm systems have the same advantages as wireless burglar alarm systems. They are easy to install, they can provide pinpoint sensor identification and location, and they are portable.

Sometimes the best way to illustrate the benefits of wireless is to draw on an extreme example. The case of a high-rise condominium complex and the fire protection required there reveals several important points about fire protection for the single-family dwelling—most notably, the benefits of a well-designed wireless fire system.



The Walnuts complex is Kansas City, Missouri's most exclusive condominium address. Given the unique history and the elegance of the three 10-story high-rise buildings comprising the Walnuts complex, it's no wonder it attracts some of KC's most affluent residents. Built in the 1930s, the Walnuts was designed with a look to the past (it was based on a nineteenth-century British model) and a look to the future. With 9 inches of concrete between each floor, the Walnuts has shifted less than a quarter inch in 65 years and is going to be standing for a long time.

The same could not be said, however, for the building's aging fire system. Residents felt that the smoke detectors, pull stations, and fire bells located only in common halls were inadequate to provide effective warning of a fire. The system offered no individual zone identification, and the alarm bells were inaudible from certain areas within the units. If a fire were to start in a residence while no one was home and a smoke detector went off, none of the other residents would benefit from an early warning. Also, in the first-floor hallways there was no protection at all, except for right at the elevator.

A system upgrade was important to the Walnuts for another reason: market competition. When a new luxury high-rise condominium complex goes up and offers sophisticated security and fire protection, it forces buyers to carefully consider the value of their investment. The president of the Executive Board of Managers at the Walnuts was well aware of the power of market forces. As former CEO of Butler Manufacturing, George Dillon knew that keeping the Walnuts competitive was a critical part of the Board's mission. "For a unit that was built in 1930 and wants to be competitive with the latest complexes, a new system was fundamental," he said. So Mr. Dillon and the Board embarked on a three-and-a-half-year process that resulted in the installation of a wireless fire and security system.

Michael Benedict, General Manager of AAA Security Systems, Inc., in Lenexa, Kansas, designed the comprehensive wireless fire system to replace the old one and added first-floor wireless perimeter intrusion protection. The system included wireless smoke detectors in common hallways and fire pull stations in the front and rear elevator areas of each floor. Each residence received a wireless heat detector in the kitchen area and wireless smoke detectors in hallway areas that lead to bedrooms. A flush-mount speaker was located in each unit as part of an annunciation system, which is activated by burglary and fire control panels. Benedict also designed and installed wireless perimeter security and a closed circuit digital system to monitor the exterior of the building. Perimeter shock and door/window sensors secure the first floor and basement windows. In addition, all doormen are equipped with wireless panic buttons.



According to Benedict, there are over 430 pieces of wireless equipment in the Walnuts. The result of a years-long process of research, consulting, bidding, installation, and cutover is an expansive system that covers more than 50 residential units over 30 floors.

Despite the expansiveness of the Walnuts installation, you should know that the process of recommending wireless was not all smooth sailing. "The first time I proposed a wireless design to the Executive Board of Managers, the room became very quiet," said Michael Benedict. "The Board, the building management people, and one of their security directors were there, and when I began to justify the wireless design, there was a lot of pen tapping, leg crossing, and shuffling of feet." It turns out the Board had already sought bids from two other companies who had said that a wireless system was impossible. "The Board was certainly polite," Michael said. "They thanked me—then I left."

Benedict had initially been asked to serve as a consultant to the Walnuts, but once he saw the lay of the land, he knew he wanted a chance to bid on a wireless installation himself. He knew it was a gamble to go for the bid and pass up the sure-thing consulting job, but he had declined the consulting offer anyway so he could set to work on a wireless proposal. After his presentation, it looked as if his gamble hadn't paid off.

But six months later the Board called him back. They had gone ahead and hired a consultant to assist in a system design, but the consultant—who recommended a hardwire system—hit several walls with the administrators on the system he had designed. The Board received hardwire quotes from two national security companies, and bids were coming in at $300,000-plus. The bids did not include the repair of floors, walls, and rare moldings, the cost of which building superintendent Larry Davidson estimated to be in the millions of dollars.

After reviewing the quotes, George Dillon and the Board decided to ask Benedict and AAA back. This time, Michael had their attention. "[Michael's proposal] was ideal for us. It was exactly what we wanted from day one," Mr. Davidson said, "but we were always told wireless couldn't be made to work." The buyers were not only happy to go wireless, they were also happy to save between 40% and 50% in labor and materials by going with Benedict's wireless proposal.



Two factors weighed heavily into Benedict's decision to stick with wireless in the first place. One was the buildings' ornate interior and exterior architectural design, coupled with the spare-no-expense construction (figure 8-1). The other was the fact that residents entertain often and have visitors like the Duchess of York and other high-profile personalities. They insisted on the very minimum of any type of interruption to their lifestyle, which translated into AAA working on site from 8 AM to 4:30 PM only—no evening or weekend work and no overtime to meet deadlines.

Benedict brought in engineers from two wireless manufacturers and submitted both companies to the exact same signal strength and performance tests. Both systems were practically identical in terms of signal strength, but he needed a system that could be operated by persons of various skill levels. Superintendent Davidson felt the system should be operable in two to three steps, rather than five or more, in order for all personnel and residents to be able to use it.

■ **8-1** *Building owners commonly select wireless systems in order to protect structural artifacts from being damaged during installation.*

Because the Walnuts is considered a commercial building, Kansas City fire marshals and codes demanded close scrutiny of the design and equipment specs before AAA could turn a screw. With some discussion and the provision of support documentation, they were issued the permit and proceeded with the installation.

The staff at the Walnuts wanted to be able to monitor the system internally. Because they have a security control room that is constantly attended, they could install a central station receiver in their existing security office.

Critical to the wireless installation was taking as much time as required to do comprehensive field signal strength tests, especially in a project this size, before beginning installation. Testing included the range from 300 to 700 feet through walls and floors to the control panels located in the stairways. AAA installed a total of nine control panels for the fire system and six additional control panels to provide first-floor perimeter protection.

"Having [devoted] my life to [dealing] with both the manufacturer and the dealer, I knew what quality standards I was looking for," George Dillon said. He wanted to know about AAA and the equipment manufacturer, what size they were, how long they'd been in business, whether they were financially sound, and where they stood vis-a-vis their competition. He wanted to be assured that the manufacturer provided the system and technological competence for the job—and more importantly, that dealer and manufacturer would be in business five years down the road. Mr. Dillon even asked what would happen in the event of a merger.

(These are the same questions you might very well ask a dealer who proposes a system for your home. Keep in mind that you want a long-term relationship with the company to whom you are entrusting your home's security.)

Aesthetics was an ever-present concern for residents at the Walnuts, as it no doubt is in your own home. Residents didn't like the looks of the red fire pull stations in the foyers of each building, so Michael moved them just around the corner. Some residents didn't like the initial location of smoke sensors, so Michael moved them as well. In short, Michael never said "no." The wireless sensors made it easy for him to comply with the residents' wishes. He could satisfy his customers and still meet code without affecting costs.

With Mr. Dillon's expertise in the building professions, he had high expectations of the installer and manufacturer, but he was also aware of the delays, missteps, and revisions that can come with any installation project. Mr. Dillon pointed out that many Walnuts residents are elderly or may have hearing or sight impairments and that they have fears about intrusion into their homes for the installation. When residents raised concerns about testing and installing, Mr. Dillon was able to explain the process and the rationale for each step. So, although his participation placed certain demands on AAA, it also freed the installers to do their jobs with increased understanding from residents.

"I even had residents express surprise when we said we were finished with their installation so quickly," Michael said. "We only had to install the two sensors and an annunciator in each unit, which really didn't take but a few minutes each." It helped that sensors he installed didn't add programming time to the installer's stay in each unit. For the control panels to learn the individual identities of smoke and heat sensors, the installer had only to trip the tamper switch in each sensor.

It also helped that Benedict prepared an eight-page safety manual for each resident, and each new resident receives one as part of their orientation. In the manual he provides a system overview and operating instructions, along with an evacuation plan.



Your own needs are no doubt a great deal smaller, yet certain principles of the Walnuts application must be observed if a fire system is to provide adequate detection and warning in a single-family dwelling. For example, sensors must be placed where they are best suited to detect smoke or fire. Here, again, the wireless advantage stands out, since sensors can be placed where they are most needed instead of where it is the most convenient to run wires.

## In the next chapter

In Chapter 9, we'll look at one of the most exciting areas of security technology today: home automation microprocessors. These state-of-the-art units can perform many functions and control several appliances and systems simultaneously.

# 9

# Home automation and wireless technology

*Home automation* is the use of microprocessors to control or perform specific functions of the home. The control or integration of security systems, lighting, appliances, computers, heating and cooling systems, and audio/video systems are among the most commonly integrated home automation functions. A home automation system normally takes the form of a central control unit and several user interfaces such as touchpads, handheld keypads, panic buttons, TV screens, computers, or telephones.

A home's electronics communicate with each other over a residential network of one kind or another. A network might be wireless infrared or radio frequency network, or a wired network. Wired communications link components to each other in such a way that a dedicated wire would connect a bank of lights, for example, in the living room to the home automation control system.

Hardwiring is generally considered the best option for new construction, since the proper cables can be installed before the walls go up. But in retrofit situations, it can be very costly to run wires through floors and walls, so wireless or "softwiring" is used. *Softwiring* uses the wires that are already in place as part of the electrical system. Softwire communications has improved a great deal in the last few years and can support very elaborate home automation networks. However, in some cases softwiring can be less reliable than hardwiring because several devices share the same wire, which can cause noise on the line.

## How home automation works

A home automation system receives signals from electronic devices and responds by sending signals to other devices. When security is part of a home automation system, a motion sensor might not only signal that an intruder is on the premises but might also signal for house lights to turn on. A security system that is operable from a Touch-Tone telephone and is also equipped with a temperature control module gives the security system owner the ability to change temperature settings over the phone from almost any remote location.

Since security is often the first step homeowners take toward automating their homes, many security system manufacturers have developed ways of making their security system controllers interoperable with lights, heating/air conditioning, audio/video, and other common household functions.

Security system controllers are commonly designed to control lights or appliances that are plugged into appropriate modules. By being connected to the security system, the lights and appliances become controllable via security system touchpads or Touch-Tone phones. Depending on the system, the user is able to turn the networked lights on or off at the same time or have individual light control. Perhaps most importantly, system lighting flashes when an intruder trips an alarm. The feature is designed to scare the intruder away and to attract attention to the home. Flashing lights also work as a warning to the homeowner that an intrusion has occurred and that he or she should not go into the house until authorities have investigated.



Some systems have an energy control module that takes over for the thermostat when directed. The module can be controlled from a remote Touch-Tone phone and used to set temperatures higher or lower than the thermostat reading. On a winter day, you can call from the office to raise the temperature so the house is warmed up by the time you arrive home. Similarly, you might let the temperature rise during a hot summer day but can call up the system to have the air conditioning come on to cool off the place by the time you get home.

Many utilities have brought automation into the home via meter-reading systems that automatically take meter readings and call the readings in to the utility. Such devices take away the need for intrusive visits by meter-reading personnel.

## Standards in the industry

The simplification that so many people enjoy from home automation comes via centralization and integration of equipment controllers. Efficiency is a result of electronic timers that turn equipment on and off to prevent wasted energy. Centralization comes via the design of devices that can integrate controls in a single unit. Although no single industry standard has yet prevailed, the 10-year development of the Consumer Electronics Bus or CEBus standard has made it possible for product manufacturers to design equipment that is interoperable with other products. Basically, the CEBus standard is an engineering design specification. More than 400 companies have participated in drawing up CEBus specifications.

Another standard platform for information distribution and control capabilities is called LonWorks, developed by Echelon Corporation of Palo Alto, California. LonWorks has grown out of industrial and commercial standards and is under the control of a single company.

## Controllers

Home automation controllers usually fall into the keypad, touchscreen, telephone, and remote control categories for control of lighting, security, closed circuit video surveillance, heating and air conditioning, audio/video entertainment systems, telephone, and intercom.

Consumer magazines such as *Electronic House* are excellent resources for finding the latest in home automation products. The February 1996 issue lists eight system controllers, some of which work with wireless security systems. Some products, such as the VuFone "smart" phone, simplify the automation of security, HVAC (heating, ventilating, and air conditioning), and lights.



### VuFone

The VuFone has an ATM-like menu. Onscreen prompts explain how to turn the security system on or off, and control the temperature settings for an energy saver module. It also controls lights and appliances. Onscreen prompts make it easy to write scripts for the VuFone that automate according to your particular schedules. The VuFone allows up to 15 different combinations of security, temperature, and light/device control instructions. For example, "Weekday Mornings" might be a script for brewing coffee by 6:30; pumping soft, classical music via a Digital Satellite System into the living room; turning bedroom lights on and motion detectors off; and setting the temperature to 72°F. (Within each script, up to four different temperature settings can be scheduled.)

### X-10

X-10 is among the oldest home automation equipment manufacturers in the country—and one of the fastest growing. Since 1990 X-10 has become the highest-volume producer of consumer-installed security systems. The key to their home automation success is their power line control units that control lights and appliances via X-10 lamp and wall-switch modules (referred to as "softwiring," above). Instead of requiring new wires run to everything you want to control, X-10 uses the network of wires already running to every room in the house for lights and appliances.

X-10 recently announced two new consoles with built-in communicators for monitoring a security system and a medical alarm system. The units are programmable by phone from a service center, with information including account number, exit, entry, and dialing delays, and timed control of lights. The Monitor Plus system from X-10 is a do-it-yourself supervised system (it reports battery failure and the location of any problem sensor) that is compatible with the full range of X-10 home automation products.

### AMP

AMP is one company that has developed a home automation controller that is CEBus compatible. Called SmartONE, it transmits control signals over radio frequency, infrared, and the home's existing power line to security, energy management, entertainment, lighting, and communications systems. AMP has also entered into a strategic alliance to introduce an integrated security, energy management, and safety lighting control system to the home management market. It's called the OnQ-mand home management product line and gives builders and developers an integrated way to take advantage of the demand for home management. The OnQ-mand system responds to a complete line of security sensors and comes with the central monitoring systems link built in. The homeowner can operate the entire system—security, energy management, and lighting—with the keypad of almost any Touch-Tone phone.



## Selecting a home automation dealer

Home automation system technology is evolving rapidly. Home automation standards are under development, and new products are being introduced all the time. Meanwhile, home systems that were once thought of as luxuries—especially security systems—are becoming commonplace. Your best course of action is to find a home/security systems dealer that you like and trust—a company whose salespeople and installers are knowledgeable and patient with your questions about the technology and its applications in your home. You'll be better off if you remember that you're shopping for more than a lot of electronic gadgetry. You're shopping for a commitment to service and information on a long-term basis.

In fact, you might decide to have one company install your security and another install other home systems. The industry is still fragmented enough that you might not be able to find one company that offers all the home systems you want. But in any event, it's likely your home systems needs or interests will change over time. Ask yourself if the company you're looking into is one with which you want to do business for service, maintenance, and additional equipment for the next 10 to 20 years.

## Resources

AMP
(800) 321-2343

Custom Command
(305) 731-0001

Home Automation, Inc.
(504) 833-7256

Interactive Technologies, Inc. (ITI)
(612) 777-2690

NetMedia
(520) 544-4567

Proteq-X USA
(800) 405-6066

Secant
(514) 935-3069

TronArch
(708) 416-6600

For other home automation resources, contact:

Parks Associates
(800) 727-5711

Advanced Services, Inc.
(800) 263-8608

EH Annual Resource Guide
(800) 375-8015

Home Automation Association
(202) 223-9669

Custom Electronic Design & Installation Association
(800) CEDIA-30

## In the next chapter

In Chapter 10, we'll take a closer look at central monitoring stations. We'll talk about the different types of monitoring available and what you can expect from a professional monitoring service.

# 10

# Central monitoring stations

Alarm monitoring has evolved significantly since the days when the only way to call for help was to call, "Help!" However, responding quickly to emergencies is hardly a modern concept. Consider, for instance, the seventeenth-century practice of setting up "rattle watches" made up of patrolmen who searched for fires at night and called for bucket brigades with the use of wooden rattles. At the sound of the alarm, citizens were very cooperative in ringing bells and placing lights in windows to show the way for firefighters who would otherwise be traveling in the dark. But even when the cry for help was heard by the citizenry, the difficult issue of locating the fire remained. The speed of sound being faster than modes of transportation, an alarm could be sounded and plenty of firefighters alerted with no way for them to tell where the fire was ablaze.



## Early automatic dialer systems

Methods have improved over the years to the point where a residential phone line could be linked to a police or fire station for direct calls for help. In early systems, when an alarm was activated, a recorded message notified police that a burglary was in progress. A dialer with two or more sending channels would have a second message and an automatic dialing response directed at a fire station. But those early arrangements were cumbersome for officials, and false alarms from too many residences soon led to the demise of the practice.

Automatic dialers are still in use, however. They are set up to call friends or neighbors in the event of an emergency, or they can be programmed to call a central monitoring station. The downside of automatic dialers to people's homes is that the prerecorded messages soon gain a reputation for crying wolf, and in some cases, their reliability as a source of security is soon exhausted.

## The digital age

With the advent of digital dialers that transmit messages electronically to a professionally staffed central monitoring station has come a communications solution with some staying power. Devised to provide fast, reliable reporting of emergencies to the proper authorities, central stations are staffed 24 hours a day with personnel who are trained to respond to alarms and contact appropriate authorities. Security systems communicate with central stations over the phone or via long-range radio communication.

The type and quantity of information supplied to a central monitoring station varies according to the design of the security system. In the most elaborate models, the central station receiver has a complete supervisory report of system status and operation, along with a history. When each sensor in a security system has its own unique identification code, the operator can tell, for example, that a burglar has broken into the back door and is moving through the living room toward the upstairs master bedroom. The operator will also know if a fire has started in the kitchen and in what direction it is headed based on the sequence of smoke or fire sensor alarms. Upon receiving the alarm, the station operator knows instantaneously the address of the alarm, whether it is a fire, intrusion, or medical emergency, and where to notify the proper authorities.



## Types of monitoring

Two basic types of monitoring are available. One is *proprietary monitoring*, in which monitoring equipment, whether located on the protected property or off-site, is owned by the security system owner. The other is *contract monitoring*, in which a dealer arranges with a company to monitor clients' systems. Most of the approximately 13,000 alarm dealers in the U.S. do not own their own monitoring equipment and contract with third-party service provider.

You'll have to decide whether you're more comfortable with a local monitoring station or a national one. Both can provide excellent service. Although you might feel that only a local monitoring station can be effective in summoning help and that it makes no sense to have a security system in Nebraska monitored by a central station in Florida, a professional monitoring station hundreds of miles away can be more effective than a locally based central station whose staff is poorly trained or poorly equipped.

## DIY systems

Some people buy do-it-yourself security systems for the primary reason that they don't want to have monitoring. They don't like the idea of having strangers connected via the telephone to electronic equipment that is installed on their living room, bedroom, and bathroom windows. They might choose not to have their system monitored at all and just hope the alarm siren will scare away a burglar or the fire alarm will alert everyone in time to get out and then call the fire department.

Others decide to have their automatic dialers call neighbors, friends, or relatives. The system owner programs the system to dial up to three or four numbers until it gets an answer. (Presumably, the owner asks permission of the person whose number is programmed in so the taped message about a fire or intrusion doesn't come as a total surprise.) Depending on the arrangement the system owner makes with the person dialed, the friend or neighbor will receive a call and a taped message.

Some owners of DIY systems can also arrange for professional monitoring. Kits usually include an 800 number to call to arrange shipment of a digital dialer (if it's not already included in the kit) and an application for service. One major advantage of the DIY monitoring arrangement is that they are often less expensive than professionally installed system monitoring, and they do not require a lengthy monitoring contract. Customers can pay one month at time and cancel whenever they want. In contrast, professionally installed systems can require one-, two-, three-, or even five-year monitoring contracts at about $20 per month.



## Central station monitoring companies

The *1996 Security Sales Fact Book* lists 66 companies offering central station services. Some provide monitoring services locally, and others nationally. Your system may be monitored by a company with which your security dealer has contracted for the monitoring of all of that dealer's accounts. Some independent security dealers have their own central station equipment, so a call from your security system will be received by an operator located in your local area. Smaller, locally owned central stations often provide a more personalized service.

Monthly monitoring fees may range as low as $9 per month and as high as $25. Those fees are based on the types of services offered by the monitoring station and the length of your monitoring contract. UL-listed central stations might charge higher monthly fees to cover the costs of equipment required to meet standards for power, phone line quality and security, receiving equipment, and computer systems that contain the critical information provided by your security system. The sophisticated digital receivers, phone line security systems, generators, phone systems console equipment, and back-up systems used by central station run from $6,000 to $100,000 each. Back-up systems are required to ensure high-quality service and redundancy in the event of power or equipment failures.

## Local versus remote monitoring

The question of local or remote monitoring usually boils down to the user's personal preference. Some people feel more secure with local monitoring by the dealer's company; others prefer having their systems monitored by a company with a national reach and reputation. Dealers usually stress the benefits of whichever type of monitoring they have to offer, but either way, monitoring is only a phone call away.



## Long-range and two-way voice monitoring

Central station product and service companies may specialize in any number of areas. Some provide long-range wireless monitoring, in which the call from your control panel goes out to the central station via long-range radio instead of over the phone lines. Others specialize in two-way voice monitoring, in which the central station operator can listen in on the proceedings in a protected area after an alarm has been signaled.

In a two-way voice system, speakers and microphones are installed in the home that allow the central station operator to hear what is going on in a protected area. The operator can also be heard on the premises. Upon hearing evidence of an intruder, the operator typically announces over the speaker, "You have been detected. What is your access code?" If it's a false alarm, the system owner can give the access code and cancel the alarm. If there is no reply or if the operator hears the sounds of breaking glass or other suspicious sounds, the police will be dispatched immediately.

Equally valuable when using two-way voice is the operator's ability to comfort the system user if there is a medical emergency. The operator can get medical or other information from the user and relay it to authorities while they are on their way to the scene.

## Licensing

In the states of Tennessee, Texas, New York, Florida, California, and Virginia, central station operators are required to be licensed. Whether a license is required or not, you can expect to have well-trained professional operators running the station. Organizations such as the Security Industry Association (SIA), the National Burglar and Fire Alarm Association (NBFAA), and the Central Station Alarm Association (CSAA) sponsor training programs for central station employees and managers.

## In the next chapter

Up to this point, we've concentrated on the basic components of wireless security systems. In the next chapter, we'll look at state-of-the-art advanced features, from keychain touchpads to fiberoptics. We'll also look toward the future to see where the technology is heading.

# 11 Advanced features of security systems

The development of wireless security products is one of the strongest forces behind the growth of the security industry. State-of-the-art devices such as keychain touchpads; electronic, or "magic" keys; light modules; smart phones; energy saver modules; digital recording modules; and two-way voice technology are used with wireless systems to provide convenience in addition to burglary detection.

## Keychain touchpads



Keychain touchpads provide several options for system users. Designed to fit on a keychain, in a pocket, or in a purse, the miniature touchpads can arm or disarm systems, activate police or auxiliary panic, and turn lights on or off from hundreds of feet away. In addition, depending on the system, the keychain touchpad can turn energy saver modules on or off, make lights flash on and off, and operate a garage door opener. The small touchpad makes systems very easy to use (figure 11-1).

■ **11-1** *Keychain touchpads provide simple system operation.*

Keychain touchpad may be added to a system either as a sensor or as a wireless touchpad by the control panel. When added to the sensor, a keychain touchpad can be bypassed or deleted, preventing lost or stolen devices from being used to operate the system.

## Digital recording modules

Digital recording modules are optional with some two-way voice equipment (see "Long Range and Two-Way Voice Monitoring" in Chapter 10). The recording module continually records a segment of audio, which is "frozen" and held in memory when an alarm occurs. This recording can be played back by the central station operator after receiving the alarm call.

The recording module provides a permanent record of what happened several seconds before and after the alarm has tripped. This means a central-station operator can retrieve the sounds of a splintering door or a breaking window prior to the alarm and then hear the intruders at work immediately after the break-in.

## Fiberoptics



Fiberoptic loops have been designed to link computers, VCRs, printers, and any other valuable equipment that can be moved easily in a protective network using fiberoptics. Once the fiberoptic cable is attached to the item being protected, a potential thief can only remove the item by disconnecting or cutting the fiberoptic cable—an act that sounds an alarm, notifies a central monitoring station, and supplies probable cause that a theft is in progress. The lightweight, nonintrusive fiberoptic cable can be used to secure virtually anything movable, indoors or out, making it perfect for home, office, hospital, and school. The cables are usually made to be compatible with both wireless or hardwire security systems.

The fact that some fiberoptic systems have passed testing criteria established by the Department of Energy—for cost, climatic testing, radiation testing, and tamper resistance—should tell you how efficient these types of systems are. The systems work by sending a pulse of light through one end of a cable loop and detecting the light at the other end of the loop using a light-sensing photo detector. If the light is not seen, indicating tampering with the secured materials, a radio or hardwire signal is delivered to a control panel, which relays the message to a central station receiver.

Some fiberoptic systems are resistant to environmental conditions—even nuclear radiation—and can fit and secure a variety of containers or storage cabinets, indicate attempts to tamper with the secured item, and be relatively quick and easy to install. Some systems, such as ITI's LightGard, can be included as part of a fire and burglary alarm system (figure 11-2).

■ **11-2** *Fiberoptic loops sound an alarm if they are broken, bent, or tampered with, making them effective in preventing theft of equipment that is in use.*



## Magic keys

Designers and manufacturers of wireless technology know that protection without convenience isn't very attractive. One of the newest security devices, which is as convenient as a key but has more powerful and versatile features than the old mechanical lock-and-key systems, is the programmable key (figure 11-3).

The keys consist of a small disk that stores programming information in its memory. When used with compatible wireless control panels, programmable keys can offer end users some very attractive operating options. When communicating with the panel's EPROM (erasable programmable read-only memory) chip, the programmable key is able to quickly transfer programming information to the panel. Information might range from simple data, such as an access code that allows the user to toggle the system on and off, to more complex operations, such as downloading a panel configuration.

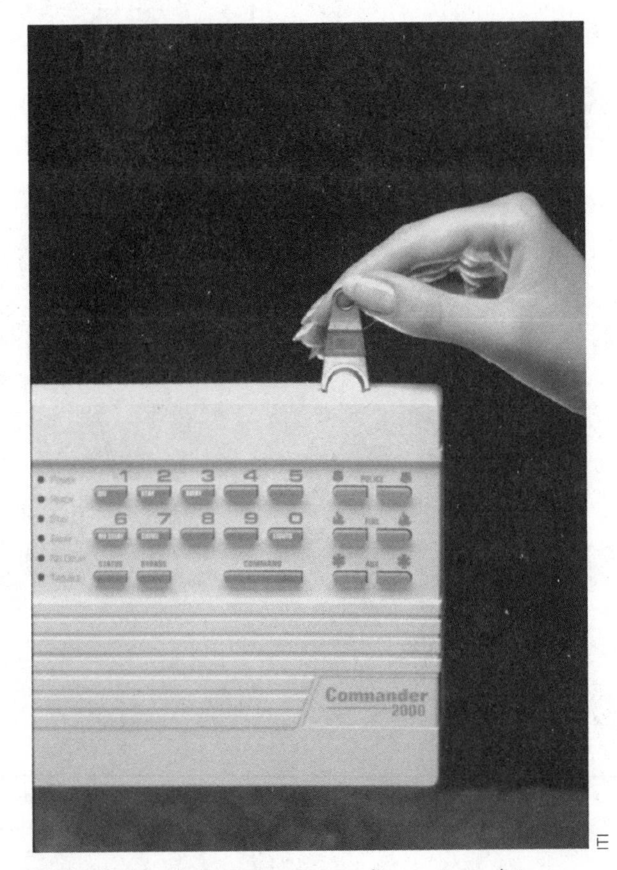


■ **11-3** *Magic keys turn security systems into access control systems.*

Keys can be programmed to allow simple but controlled access for friends, relatives, or service personnel. Access control keys allow access only on designated days. If you want a baby-sitter to be able to turn your security system on and off, you can give the key a temporary access code that will work only on the day for which the sitter is scheduled. If you're expecting a repair service Monday morning, you can provide an access key that will work on that day only.

Programmable keys are a perfect device for allowing temporary access without giving away permanent access codes. Users disarm the security system by simply inserting the key in a slot located at the top of the control panel and removing the key a few seconds later. To rearm the system the key is simply reinserted and removed once the system has rearmed itself. By using programmable keys, the system owner has not risked compromising the system by giving away a secret code.

In another application, an apartment access key allows apartment managers to give maintenance and repair personnel access to apartments only on the day for which the key is programmed. The key can also be programmed for a specific number of uses. Since the key can only be used for a designated number of times, access is dictated by job responsibility, giving supervisors and superintendents increased security control.

The uploading and downloading capabilities of programmable keys save installers large amounts of programming time. These keys serve as panel memory backup. In the uploader/downloader mode, certain panel configurations can be read by the key from the panel's EPROM and stored in the key's memory. The panel configuration can then be copied back into the panel from the key. The procedure entails putting the panel into dealer program mode, entering the download command, and inserting the key in the panel. The previously stored panel configuration can then be downloaded into the panel's EPROM when conducting, for example, a system upgrade. The process eliminates the need for the dealer to spend time adding sensors to a panel, and standard configurations such as central station phone numbers and exit/entry delay times can be downloaded to several panels without tedious, repetitive programming.



## Personal emergency response system

Home health care is becoming an attractive alternative to costly hospital care. Industry watchers predict that expenditures in home health care will double in the next 10 years. One estimate goes as high as 25% of the overall cost of health care going to home care. The shift away from hospital to home care comes as a result of many factors, including the high cost of hospital care and the growing proportion of seniors in the population. With the increased numbers have come increased concerns about the health and safety of aging parents, friends, and siblings who live without continuous contact with care providers.

One of the newest products modified by engineers of wireless security systems to meet the needs of this growing market is the personal emergency response system (figure 11-4). Designed to address the growing concerns associated with aging and with home health care, personal emergency response systems are among the most attractive new additions to dealers of wireless home security. Although wireless technology can do little to prevent persons from falling or from becoming ill, it can provide safety measures for elderly persons and some peace of mind for their friends and family.

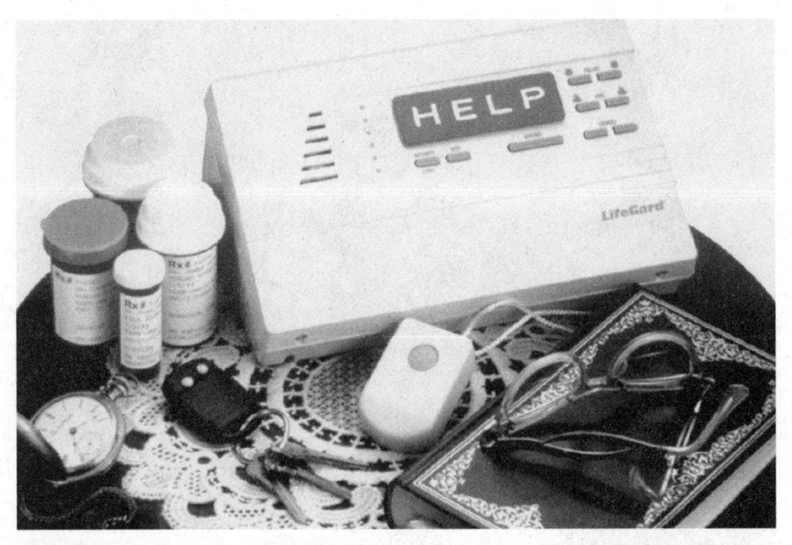


■ **11-4** *Wireless medical alert panels use security system technology to call for help in the event of an injury or medical emergency.*

A common worry among friends and relatives is that an aging parent living alone will sustain an injury and have no means of calling for help. Another is that he or she will forget to take medicine at their prescribed times. Also, as persons age, their mobility can be impaired for any number of reasons, slowing reaction time to emergencies such as fire.



Systems consist of a control panel, a siren, panic buttons, a panic pendant, and a PIR. Installation of some systems is simplified with the use of a table mounting base. Instead of finding a place to wall-mount the control panel, the unit is placed on a table or other flat surface in the living room, bedroom, or kitchen where there is access to AC power and an incoming phone line. A wireless PIR is placed to view routine traffic patterns. When the system is armed, the PIR detects no activity, signaling the panel to activate a low-volume siren for a designated period of time, for example, five minutes. If everything on the premises is all right, the user can reset the system within the delay time. But if nobody resets the panel within five minutes, that could indicate trouble. The central station is notified and authorities are dispatched. Or, if two-way voice is available, the user can be in voice contact to verify or cancel the alarm.

Wireless systems operate with wireless panic buttons that can be worn at all times, offering a physical reminder of the user's link to outside help. So-called "pill-minder" features let users know when it's time to take medication. The system can be set to beep for one minute several times per day as a reminder to take pills or to make phone calls, let the dog out, or to perform other tasks on a regular basis. Some systems also work with fire detectors and with environmental sensors that can detect furnace failure.

It's important to note that among the several panic pendants available, some are water-resistant and some aren't. Water-resistant models allow the user to wear the panic pendant in the shower, where they might be most vulnerable to injury from a fall. Some pendant batteries can be replaced by the owner without damaging water resistance, and some can't.

## Buddy system

The buddy system links the control panels of two separate systems together. It provides neighborhoods crime-watch protection without monthly meetings, turns your neighbor into an ally, and is one of the best reasons to takes advantage of wireless technology. The system offers insurance against a burglar intent on cutting phone lines, but it doesn't add hundreds of dollars to the cost of security. Designed with care, a buddy system is an excellent tool for deterring crime via detection.



The buddy system is among the most affordable options for protecting phone lines. Basically, the buddy system costs the price of a transmitter. In areas where phone companies don't offer a cut-line detection service, the cost of a $90 transmitter is far below the cost of a protector box or the use of a cellular phone. As an added benefit, it requires that you make contact with a neighbor, which in itself is a worthwhile endeavor because it creates goodwill as neighbors unite behind a common concern for security.

Some security dealers are happy to give away the extra sensor it takes to rig a buddy system when it means installing systems in two homes instead of one.

## Table mounting base for control panel

Table mounting bases for system control panels allow quick and easy installation of control panel, siren, and alarm verification/recording modules in one integrated unit. Many do-it-yourself systems are of the plug-and-play variety, but some professionally installed systems have adapted their designs to make them easier to install as well.

The table mounting base makes installation as easy as plugging in a phone. By eliminating the need to drill holes for wall mounting, the table mounting base also allows a quick and clean installation. Installation is so easy that a salesperson could install an entire temporary security system in 10 to 15 minutes during a sales presentation.

With the base, a control panel can be placed on the table next to a bed, on an office bookcase, or on the kitchen counter, making it very accessible. The base also makes it possible to rent temporary systems during vacations or to protect a vacant home that is for sale.

A table mounting base might also have room for additional components, such as a siren or a two-way voice module, which allows the central station to have a two-way conversation with anyone in the protected area. The module enables an operator to verify actual alarms and cancel accidental ones.

## Repeaters

Repeaters extend the range and power of signals transmitted by sensors. Designed to work with sensors located on the fringe of panel reception, they help to eliminate supervisory conditions in systems where the installation environment affects signal strength. A "dumb" repeater can repeat signals from sensors; a "smart" repeater can retransmit signals from other repeaters and from sensors.



The repeater's range adds power and flexibility to wireless installations. Repeaters commonly have tamper switches that cause an alarm if the repeater's cover is removed. Once programmed into panel memory, the repeater sends a supervisory transmission every 64 minutes.

## Ceiling-mount PIRs

A ceiling-mount wireless PIR can continually survey a room, hallway, garage, warehouse, and areas where a wall-mount PIR could be tampered with or obstructed. Because it mounts on a ceiling instead of a wall, the ceiling mount is especially attractive to schools, colleges, nurseries, churches, libraries, and other others where inquisitive fingers can cause tampering problems. Warehouses and other storage areas where goods are stacked need a ceiling-mounted PIR to prevent the signal from being obstructed.

The ceiling-mount PIR gives a 360° overview of the protected space. Like any PIR, the ceiling mount reacts to the infrared radiation given off by anyone moving into the protected area. The sensor's selective pulse count feature allows adjustment of its sensitivity, and included with the detector is a masking kit for blocking out common sources of rapid heating, such as forced-air ducts. Depending on area configuration, the sensor can monitor up to 1250 square feet. The ceiling mount can work alone or in tandem with a wall-mounted PIR.

Some ceiling-mount PIRs are powered by a long-life lithium battery that will provide three to four years of service under normal conditions. Operation temperature range is 10 to 120°F.

## Glass-mount sensors

Wireless glass-mount sensors are perfect for homes with large picture windows and for businesses with large storefront glass. Since the windows are stationary, no switch is necessary to detect openings. Instead, glass-mount circuitry analyzes and detects the frequency of breaking glass. The sensor is mounted on the glass to provide dependable security for home and business.

## Long-Life sensors



In 1996, ITI announced the development of a new, long-life door/window sensor with an expected transmitter battery life of 20 years in typical residential use. The Long-Life sensor/transmitter provides the same range and power of other security system sensors, but battery life is three to four times longer.

This is good news for the wireless security industry. Developed to reassure consumers of the convenience of wireless systems and to reduce dealer service calls, the sensor represents an important advance in wireless security's appeal. Consumers do not want to pay for a false sense of security, and they are often wary of systems relying on battery power. If a battery-powered sensor ceases functioning because of a dead battery, the system is not providing full protection against undetected entry.

In addition, many consumers are apprehensive about changing sensor batteries, fearing they might set off a false alarm. This sensor is a step in the right direction, aiming to forestall objections to equipment design that can contribute to unreliable performance. To increase the sensor's reliability, its lithium battery is soldered into place. Instead of replacing the battery 20 years down the road, the entire sensor will be replaced. According to ITI engineers, the

20-year battery will outlive current sensor technology, eliminating battery life as a consumer issue altogether.

## Lithium-powered heat sensors

As security systems and their advanced features grow in popularity, manufacturers respond to the increased demand by building new features into components. The learn mode rate-of-rise heat sensor with lithium batteries is a case in point. Once seen as a luxury item, the heat sensor is now so commonly installed that it has been modified to use a lithium battery. Under normal conditions the battery in the heat sensor will last at least four years. The heat sensor initiates an alarm when a fixed temperature has been reached. The sensor detects the rate at which the temperature is rising and responds to an increase of 15° per minute or greater.

## Panic pendants

If a personal panic button gets misplaced before or during an emergency, it's not much use, so manufacturers are making new pendant transmitters that are almost impossible to lose. Using a soft breakaway neck cord allows the user to safely keep the water-resistant pendant close at hand, even in the tub or shower. A belt clip keeps the pendant close by, whether the user is in the garden or the garage. A wall-mount clip enables mounting in the bedroom, near the tub, or next to a favorite chair.



The range of panic pendants varies from brand to brand. Keep the size and layout of your property in mind when choosing a system. Pendants typically transmit 1000 feet or more in open air, which means a pendant can be used upstairs or downstairs, inside or outside. The size and power of panic pendants make them ideal for times when the user is home alone.

Long-life batteries for the pendant typically last five to eight years. Some pendants have to be replaced once the battery wears out. Others are designed to have their batteries replaced by the system user. Check before you buy to make sure a pendant's battery can be replaced without it losing its water resistance.

## Smart phones

Smart phones can do all kinds of tasks. They can read credit cards and provide the latest stock reports; they can link you with interactive services provided by phone companies and retail outlets. Smart phones can also be used as an integrated command center to control security systems, lighting, and heating. As part of an integrated comprehensive home automation system, a smart phone can also give you access to electronic home banking and shopping.

Smart phones simplify telecommunications options. A menu gives users fingertip control over arming and disarming the security system, and it controls lights and heat, panic buttons, and latchkey functions with the ease of an ATM. Some smart phones have advanced phone features such as Call Forwarding, Call Waiting, and conference calling built in. Additional business alliances expand opportunities to use smart phones to order products from home or to pay bills, check bank balances, transfer funds, and buy stocks.

Smart phones with built-in computer keyboards, advanced speaker phones, and electronic directories make short work of searches for addresses and phone numbers, and getting a conference call off the ground is as easy as making a standard phone call.

## Telephone control



Being able to communicate with your security system via the telephone is a great convenience; you can control the system from work, school, or anywhere there's a Touch-Tone phone. If your plans change after you leave the house, you can call up your system and change the temperature without having to go back home.

## Light control

Light control is part of all sophisticated security systems. Lighting for security is as important as locking the doors at night. Advanced lighting control gives you automated dimming and can duplicate random patterns of lights going on and off, imitating the patterns of people when they are home. Some security systems automatically flash lights in the house when an alarm has been triggered and continue to flash light to indicate that an intruder has been in the house. If the user should drive up and see house lights flashing, he or she would contact police before entering the house, since the intruder might still be inside.

## Temperature control

Controlling energy consumption is one of the best ways for a household to save money. By automating temperature control via your security system, you can shave large percentages off your utility bill. The security system interfaces with a set-back thermostat that automatically supersedes the regular home thermostat. These thermostats can work with both air-conditioning and heating systems.

## Meter reading

Automated meter reading is quickly becoming commonplace. Readings are often transmitted over the phone lines to your utility in the same way as security information is transmitted to your central monitoring station. You might already have been approached by your local utility about their new security offerings. Some people find a great advantage in having their utility install their security system, as it eliminates the need to go through the buying process with an unknown security dealer.

## Environmental sensors



Temperature gauges linked to door/window transmitters can send an alert in the event of furnace failure, which could lead to frozen pipes. Increases in temperature can also be dangerous, such as in poultry farms where overheating can cost thousands in livestock losses. Wireless transmitters are used in sewage treatment plants to warn of dangerous amounts of toxic chemicals or rising levels of effluent.

Carbon monoxide detectors are becoming more common all the time with increases in reports of death caused by the toxic gas. Some wireless sensors can be bought as an isolated component and used separately from a security system. They can either be plugged into a wall outlet or are powered by batteries. In a supervised system they would report to the security system control panel.

## In the next chapter

Chapter 12 discusses how a professional alarm company goes about planning the installation of a wireless installation. Knowing how the pros do it will help you select a qualified installer or help you plan your own DIY system.

# 12 How the pros plan a wireless installation

Before you meet with a security consultant in your home, you should have an idea of what goes into planning an installation. Doing so will allow you to ask informed questions and participate in designing a system to match the needs of your family and of your house. This chapter describes the planning process a professional dealer/installer follows for a wireless installation. Some of the items on the planning checklist below are steps a professional installer would discuss with you in preparation for a successful installation. For illustration purposes, we are using a system with several advanced features.

When planning any installation, the installer must do the following:

- ☐ Determine the purpose of the system.
- ☐ Plan hardwire sirens and piezos.
- ☐ Plan the use of wireless components.
- ☐ Determine component locations.
- ☐ Plan wireless interior sirens (WIS) and X-10 lamp modules (if included in your system).
- ☐ Plan to explain the system to all users of the system.

A note before getting on with the main focus of the chapter: systems are available for lease or for sale. You need to know whether you want to invest in a system that you will have for the rest of its life or to lease a system that goes back to the leasing company after a designated period of time. As in all choices of this kind, cost usually comes into play. Leases provide a lower monthly payment but no ownership. Buying a system might cost you more up front or on a monthly basis, but in the end, you own the system. It's a basic part of the process to understand whether you are signing a lease or a purchase agreement for your security system, and you must tell your security dealer whether you want to buy or lease.

## Determine the purpose of the system

The security system used in this example can be used as a fire warning system, an intrusion alarm system, an emergency notification system, or any combination of the three.

Following are the components for your system, with fire warning, burglar alarm, and signaling equipment, that you may add according to your needs and budget.

### The basic system

- ☐ Control panel (controls the system)
- ☐ Line carrier power transformer (uses household current to power the panel and wireless interior sirens).

### Household fire warning system

- ☐ Smoke sensors and heat sensors added to the basic system

### Household burglar alarm system

- ☐ Door/window sensors, motion sensors, sound sensors, shock/glass-break sensors added to the basic system



### Home health care signaling equipment

- ☐ Water-resistant panic pendant

## Plan hardwire sirens and piezos

Sirens produce alarm sounds; they are set off by armed sensors in your security system. Piezos produce status sounds (to indicate the status of the system) in areas of the premises where the panel speaker cannot be heard. For easiest operation, piezos should be installed so that you can hear status beeps from any system touchpad in the house. That way, you will always know at what level your system is set. To determine the location for remote sirens and piezos, test the range of the panel speaker. You could have a friend change arming levels of the system while you place yourself at various locations throughout the house.

## Plan the use of wireless components

Depending on the size of your house, you will have a control panel that can respond to a certain number of system sensors. On some do-it-yourself kits, you can add an unlimited number of sensors to the system without any problem because they are not supervised and only have to report alarms to the control panel. Keep in mind that adding sensors to a supervised system becomes increasingly restrictive because of the amounts of memory available to the system for supporting the advanced features.

As you plan, keep track of the type and number of wireless sensors required for the installation (see planning forms in the Appendix). During the installation process, each sensor will be programmed into a group. The group number identifies the sensor as belonging to an intrusion detector, fire detector, or medical emergency device. Code numbers provide the control panel with the information it needs to operate according to the needs of your home.

Since the sensor's group number identifies the sensor's purpose as intrusion detection, emergency panic button, or fire detection, the central station operator will know the purpose of each sensor during an alarm. The group assignment for each sensor will be indicated in the installation instructions of your security system. Keep track of the group numbers you have assigned.

When planning where to locate a wireless transmitter, first test to see that the transmitter is within range of the receiver. To test transmitter range, follow these steps:



1. Enter a sensor test code.
2. Place the sensor where you want it to be located.
3. Trip the sensor.
4. Note the number of status beeps.

## Determine component locations

Follow these guidelines for locating components:

- ☐ Provide the panel with access to the incoming phone line, to 110-Vac power, and to other wired devices.
- ☐ Make sure you can run the necessary wires between the panel location you select and the locations of the hardwired components and connections.
- ☐ Mount the panel in a temperature- and humidity-controlled environment.
- ☐ Mount sensors within 100 feet of the panel whenever possible. Although the system has an open field range of at least 500 feet, use 100 feet as a starting point inside a building.
- ☐ Refer to detailed mounting instructions that come with each sensor.

The general guidelines for locating sensors are as follows:

**Door/window sensors** For optimum performance, it's best to locate the sensor on the door frame and the magnet on the door. This prevents the sensor from being jarred as often as it would if it were on a swinging door. On a double-hung window, place the magnet on the bottom movable window and the transmitter on the stationary upper window. For recessed door/window sensors, drill the transmitter hole into the door frame and the recessed magnet hole into the door.

**Motion sensors** Motion sensors are the number one device for causing false alarms. One of the best ways of preventing false alarms caused by motion sensors is to install them correctly. The most common placement error is to point the lens of a motion detector toward a window where changes in temperature are sure to be picked up by the sensor. Other problem areas include heating ducts. When the heat kicks on in the house, the motion sensor detects the rise in temperature and can't distinguish the furnace heat from the body heat of an intruder. Pets can also set off false alarms.



Be sure to follow manufacturer's instructions for the correct placement and masking of motion sensors in your home. (*Masking* refers to creating blind spots on the sensor eye to allow for the movement of pets). They are wonderfully versatile and sensitive devices, but they must be used properly. Some homeowners use security overkill by putting door/window sensors on interior doors and then adding motion sensors inside the room already protected by the door/window sensor. If there's no way a person could get into the interior room without going through the door, there's no reason to put a motion sensor inside the room. Use motion sensors in long hallways or for perimeter protection in rooms that have several windows. But remember: don't point the motion sensor in the direction of a heat source. That includes windows, vents, and radiators.

**Sound sensors** Sound sensors hear specific frequencies caused by breaking glass. Sound sensors are usually placed on walls or ceilings. They are especially economical in rooms that have a lot of glass. The ideal location for some sound sensors is straight across the room from the protected area. Sound sensor may be set to detect two different frequencies to prevent false alarms caused by barking dogs or the clattering of dishes.

**Shock and glass sensors** Shock sensors are also used to detect breaking glass but are often mounted on a window frame. These sensors should only be mounted on window frames in which the window fits snugly and does not move or rattle. Shock sensors must be mounted on window frames that are larger than the sensor's base. Glass sensors are mounted right on the glass of windows that are stationary.

## Plan wireless interior sirens and X-10 lamp modules

If you will be using the wireless interior siren (WIS) or the X-10 lamp module, you must use the line carrier power (LCP) transformer to power the system. The LCP transformer allows the WIS and the X-10 module to receive signals from the control panel via household wiring. The WIS produces low-volume status sounds and high-volume alarm sounds. It does not produce voice messages. X-10 lamp modules turn on lamps during police, auxiliary/medical, and fire alarms and during entry/exit delays. The number of X-10 modules will depend on the needs or desires of your home. The number of X-10s is limited by the number of available outlets on the premises (the same is true of wireless interior sirens).

## Explain the system to all users



If you buy a system from a dealer, he or she will probably use a demo kit to show you the components and how they work. You should have a chance to press the buttons in the installer's presence to give the system a variety of commands. You should also be allowed to hear the sirens so you know what your system will sound like in actual use. Trying out a handheld panic button or pendant lets you know what it feels like to activate the system with a simple press of a button. A dealer who uses a demo kit is selling and teaching at the same time, which works out well for everyone. In addition, be sure to peruse the owner's manual while the installer is still there in case you have any questions.

Some companies offer a video owner's manual. If one is available, be sure to watch it. Watching the video gives you a better understanding of the system's basic functions and gives you a chance to formulate questions that a dealer should be happy to answer for you.

Instructing all users on the operation of the system is important because the more all system users know, the fewer false alarms you'll have, and the fewer panic calls made on simple matters like how to turn off a low-battery message. Plan to cover the following topics in your training:

- ☐ How to prevent false alarms
- ☐ How to cancel an alarm
- ☐ What siren/piezo sounds the system makes and what they mean
- ☐ When to call for system service
- ☐ How to command the system using a phone on and off premises (with applicable systems)
- ☐ How to test the system

## In the next chapter

Now that we understand the rudiments of planning a security system, in the next chapter we'll dive into the details and describe exactly how a system is installed.

# 13 What to expect during a wireless installation

Depending on the size of your home and the number of sensors you have installed, the installation of a wireless security system by a professional security dealer can take from four to eight hours. The installation itself will be fairly straightforward, as the planning has already been completed by the security consultant who walked through your home and worked out the number and location of the system components with you. The installer's job is to follow the instructions that you and the consultant have already worked out. The installation usually includes the following steps:

1. Mounting the control panel
2. Connecting the phone cord to the panel
3. Connecting the panel to the incoming phone line
4. Wiring the line carrier transformer
5. Programming the panel and sensors
6. Programming and installing wireless touchpads
7. Connecting the transformer to the panel
8. Applying power to the panel
9. Testing the system
10. Troubleshooting the system

The installer will usually use the following tools:

- ☐ 22-gauge two- and four-conductor wire and wire stripper
- ☐ Phillips and flathead screwdrivers
- ☐ Drill and a variety of small bits
- ☐ Digital voltmeter
- ☐ Weatherproof crimp connectors and crimping tool

## Mounting the control panel

During the planning phase of the installation, the security salesperson found a centralized, temperature- and humidity-controlled location on the premises with access to an incoming phone line and 110-Vac power. The location of the control panel does not always have to be centralized, as the range of many transmitters is much farther than the entire length of a house. Often the location of a convenient phone line connection and electrical power source suggests an appropriate place for the control panel.

One other variable in the installer's decision on where to put the control panel is its design. If the panel is designed with a touchpad or is a table-mount type of controller, it will be installed where it can be used conveniently and where it will have a pleasing appearance. If the panel is purely functional (these panels look like other utility boxes stuffed with wires and circuit boards), it is usually installed in a closet where it will be out of sight.

In the installation that follows we'll be discussing a moderately priced control panel called the Commander 2000. This control panel can receive either 8 or 17 wireless zones (that is, up to 8 or up to 17 individually identified sensors), which makes it a good choice for a modest home. A large home would require 60 to 90 zones and up.



The following steps are involved in mounting the Commander 2000 control panel:

1. Remove the battery door and battery bucket.
2. Hold the panel upside down against the wall at the desired height and location (figure 13-1).
3. Mark the location of the mounting holes.
4. Drill two holes.
5. Put inserts and screws in place.
6. Hang the panel right side up on the screws.
7. Mark the lower mounting holes (figure 13-2).
8. Remove the panel, and repeat Step 5.
9. Mount the panel on the four screws, and gently tighten the lower screws.

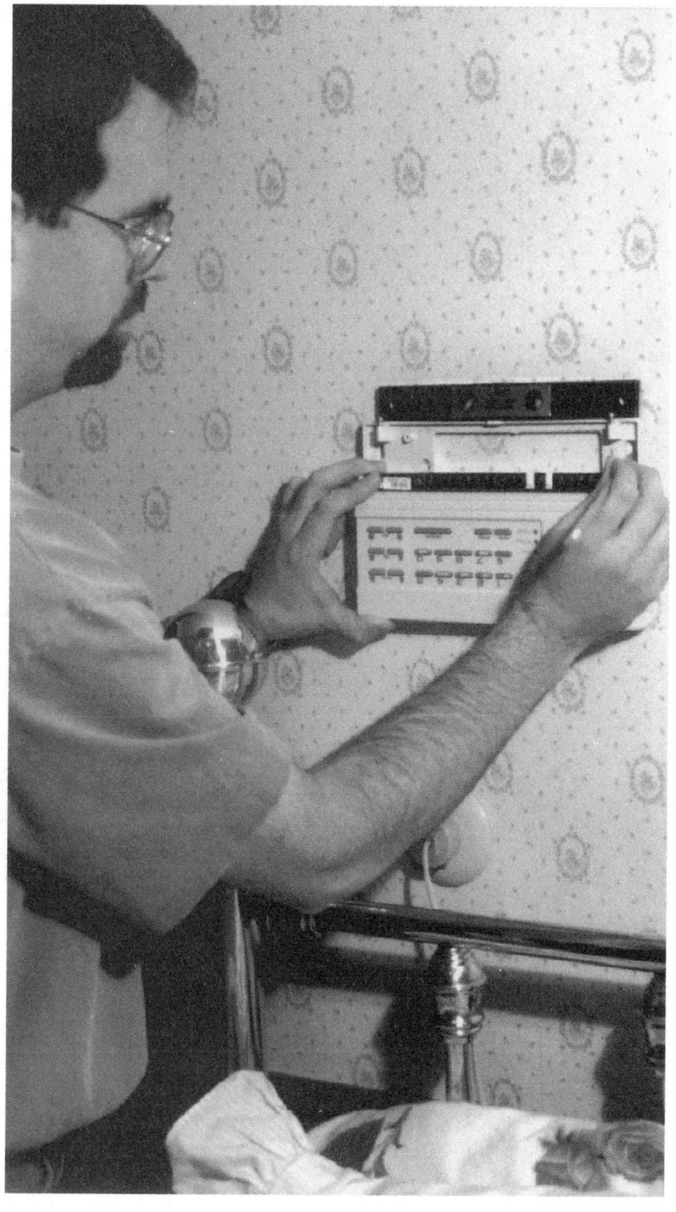

■ **13-1** *Once a location is selected, holes are drilled for wall-mounting. Note the use of a level while marking the holes.*

■ **13-2** *Marking the bottom holes for wall-mounting.*

## Connecting the phone cord to the panel

To provide optimum service, the control panel must be able to override other phones in the event of an emergency. This overriding is called *line seizure*. The DB-8 cord provided with the Commander 2000 is a phone cord with a modular plug. The modular plug connects to the RJ-31X jack, which is connected via a four-lead splice wire to the telephone company (Telco) protector block and the premises phones. The jack lets the installer unplug the panel from the phone system, if necessary, a capability required by many local ordinances.

To connect the panel to the phone line the installer does the following:

1. Connect the DB-8 cord to the panel (figure 13-3) using the following table:

| **Terminal number** | 15 | 16 | 17 | 18 |
|---|---|---|---|---|
| **Wire color** | Green | Brown | Gray | Red |

2. Wrap each end of the four extra wires with electrical tape to insulate them. Tape them together in case they are needed for future use.

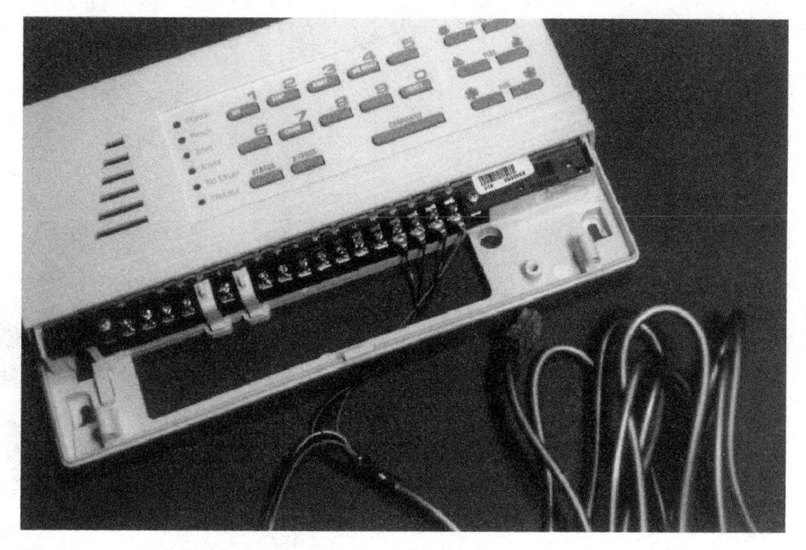

■ **13-3** *The telephone cord is attached to the control panel phone circuits.*

Now the installer checks phone line polarity. Reversed polarity somewhere in the phone system is a common cause of phone problems. Checking phone line polarity before making connections reduces the risk of such problems. In general, these are the steps your installer will follow:

1. Locate the Telco protector block where the phone line comes into the premises.
2. Identify the positive terminal on the Telco block by using a digital voltmeter that measures dc volts (figure 13-4).

Connect the positive lead of the voltmeter to one terminal of the Telco block. Connect the negative lead of the voltmeter to the other  terminal on the Telco block.

If the voltmeter displays a positive voltage, connect the positive terminal to the positive lead of the voltmeter. Mark that terminal positive (+). If the voltmeter displays a negative voltage, connect the positive terminal is connected to the negative lead of the voltmeter.

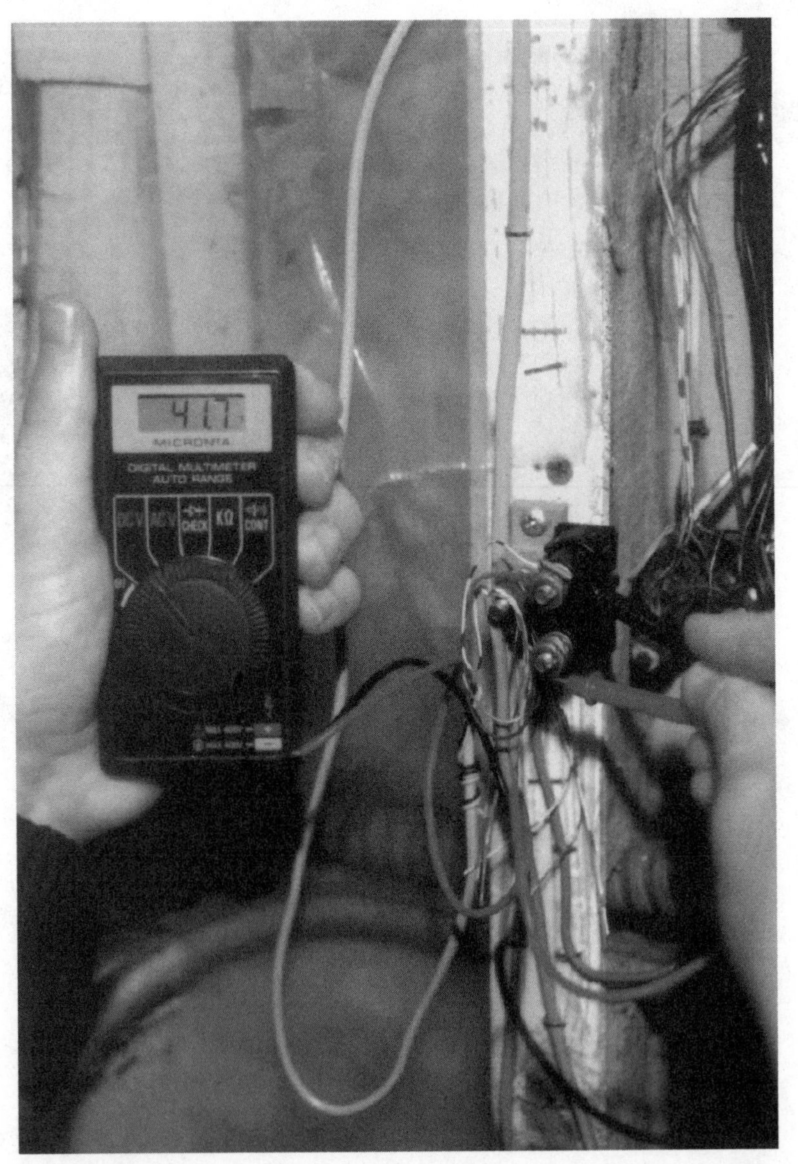


■ **13-4** *A digital voltmeter is used to determine phone line polarity.*

## Connecting the panel to the incoming phone line

A length of 22-gauge four-lead wire is used to make the connection between the RJ-31X jack and the Telco block and premises phones. The installer follows these steps:

1. Locate the RJ-31X jack (CA-38A in Canada) within reach of the DB-8 cord.
2. Run a 22-gauge four-lead splice wire from the Telco protector block to the RJ-31X jack.
3. Connect the splice wire to the RJ-31X jack as shown below (figure 13-5).

| **Splice wire** | Black | Green | Red | White |
|---|---|---|---|---|
| **RJ-31X jack** | Brown | Green | Red | Gray |



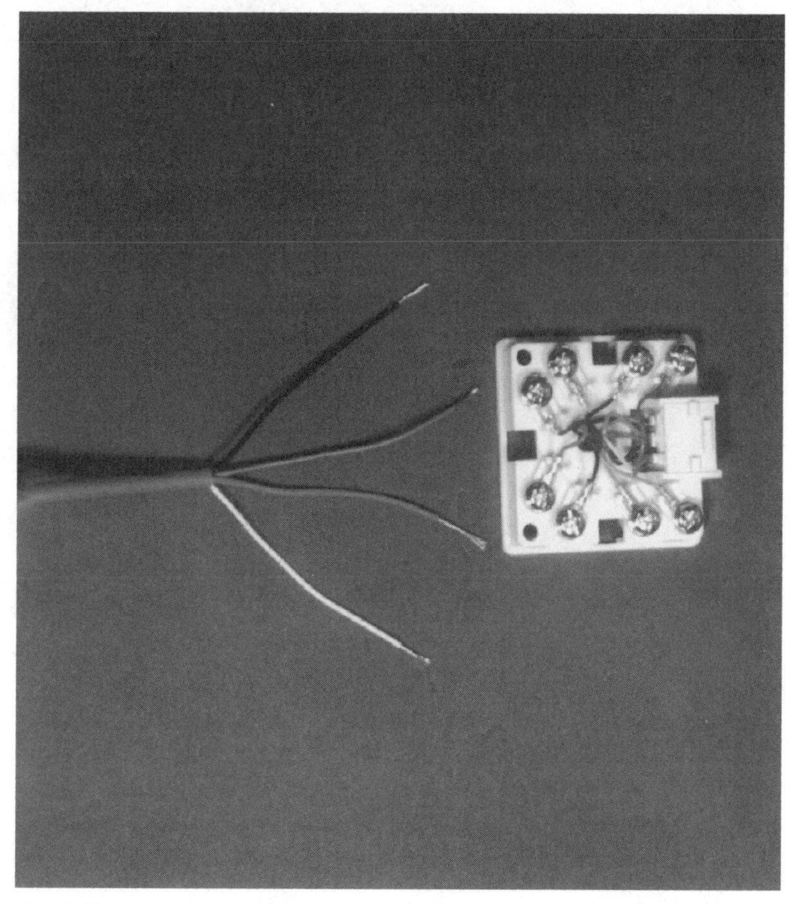

■ **13-5** *A splice wire connects the alarm panel phone jack to the telephone system.*

4. Disconnect the Telco and premises phone lines at the Telco protector block. If there are multiple phone lines, keep the positive and negative leads grouped separately when you disconnect.
5. If necessary, determine which lines are Telco lines and which are premises phone lines using a voltmeter (premises phone lines won't register any voltage; Telco lines will). See figure 13-6.

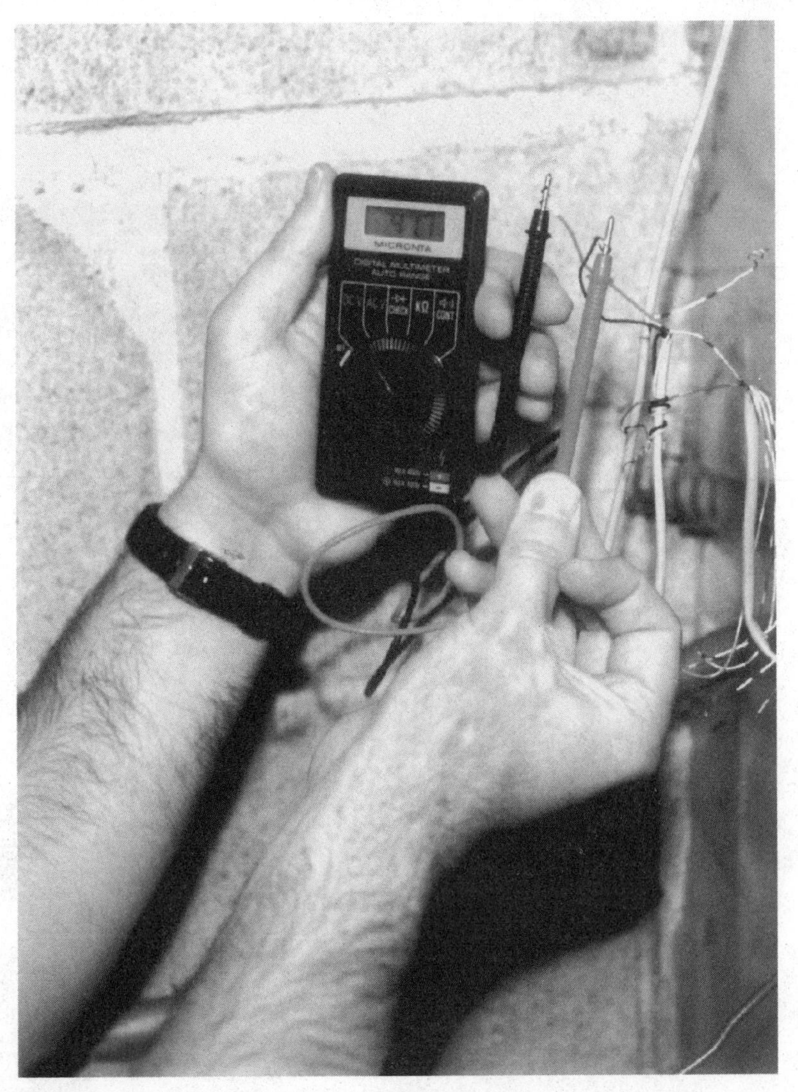


■ **13-6** *Use a digital voltmeter to determine which are the telephone company's incoming lines and which lines are connected to the home's phones.*

6. Connect the splice wire's green lead along with the Telco positive lead to the positive terminal on the Telco block, and connect the splice wire's red lead along with the Telco negative lead to the negative terminal.
7. Connect the splice wire's black lead to the premises phones' positive wires.
8. Connect the splice wire's white/yellow lead to the premises phones' negative wires.
9. Check all premises phones for dial tone and dial-out operation.
10. Plug the DB-8 cord into the RJ-31X jack.
11. Check all phones again for dial tone and dial-out operation. If the phones do not work properly, double-check polarity and wiring.

## Wiring the line carrier transformer

The line carrier power (LCP) transformer supplies dc power to the panel. This transformer also transmits line carrier signals that operate the wireless interior siren (WIS) and X-10 lamp modules.

1. Find an unswitched plug in the basement or another unobtrusive location for plugging in the transformer.

   **Note:** To avoid a possible short circuit, do not plug the transformer into the outlet at this time.
2. Connect a 22-gauge stranded four-lead wire to the line carrier transformer and to the first four terminals of the control panel (figure 13-7). Be sure to connect the same-color wire to the same terminal numbers of the transformer and the panel. (If you connect the red wire to Terminal 1 on the transformer, connect the other end of the red wire to Terminal 1 on the panel, etc.)

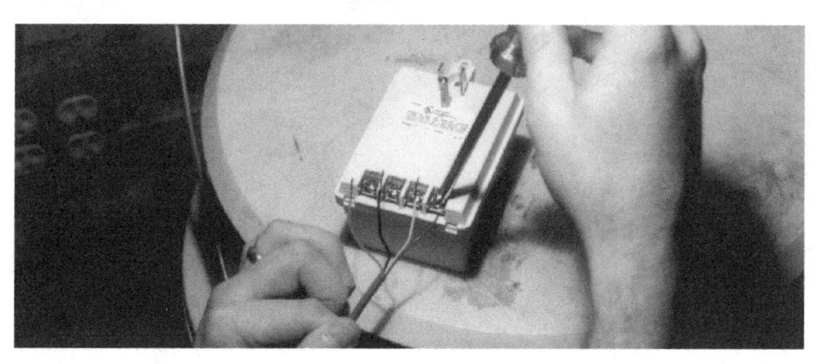

■ **13-7** *Wiring the power line carrier transformer.*

## Programming the panel and sensors

After backup batteries are properly installed, the installer will plug in the transformer to an unobtrusive plug that is not controlled by a switch. The following guidelines explain the basic steps for programming a panel. For complete programming instructions, installers attend manufacturers' training. The panel must be programmed to receive and recognize signals from sensors.

1. Enter the program mode.
2. Enter the installer programming code (assigned by manufacturer).
3. Clear panel memory.
4. Assign group numbers to all sensors.

The panel must be programmed to receive and recognize signals from sensors. Each wireless sensor has a unique sensor identification code (ID). Because each sensor ID is different, the panel has to add the ID of each sensor to the system. To enter the program mode the installer will

1. Enter the primary access code + 1.
2. Loosen the screws on the battery door until the READY light turns off.
3. Enter the installer programming code (assigned by manufacturer).
4. Identify the sensors you will install and locate the group numbers that apply to each sensor's function.
5. Record group and sensor number assignments on the Sensor Groups and Location Worksheet (see Appendix). Be sure to keep group numbers together, and assign sensor numbers sequentially within groups.
6. Use the stickers provided in the panel's accessory pack to identify each sensor when you mount them.
7. Press STATUS + [*group number*].
8. Enter the sensor number. Refer to a System Planning Worksheet for the sensor number that you planned for the system.
9. Trip the tamper switch of the sensor you are programming (usually by simply removing the cover). The panel adds the ID when you trip the switch.
10. Repeat Step 8 until the desired sensors are programmed in the current group.

11. Return to Step 7 to select a new group.
12. Press COMMAND to exit from the adding wireless sensors mode.

## Programming and installing wireless touchpads

The handheld and wireless wall-mount touchpads allow you to control the system without having to go to the panel. Most programming operations can be done with a wireless touchpad. The Commander 2000 system accepts up to four wireless touchpads.

To program the wireless touchpads an installer will follow these basic steps:

1. With the panel in program mode, press STATUS + STATUS + [*ID number of touchpad, from 1 to 4*].
2. Press BYPASS on the wireless touchpad that you want to add.
3. Repeat Steps 1 and 2 for each wireless touchpad.

The installer will refer to a planning worksheet similar to the one in the Appendix for optimum location of sensors and touchpads, and the attach sensor mounting brackets according to manufacturer's instructions.



## Testing the system

The installer will run tests to make sure all components are working properly. Testing the system occurs after installing a new system, while servicing a system, and after adding or removing devices. Testing falls into three categories:

1. Testing the sensors
2. Testing phone communication
3. Testing communication with the central station

### Testing the sensors

The sensor test lets your installer determine whether or not signals are being received by the control panel. It also tells the installer how many data rounds transmitted by each sensor were received by the panel. The goal of testing is to determine the quality of the sensor location in relation to the panel. Because the sensor test only tests sensor operation for the current installation conditions, it is important for the installer to test again if any changes are made in the environment or equipment.

To perform a sensor test the installer will:

1. Place all sensors in their secured state, normally open or normally closed.
2. Replace the battery door on the panel if the door is off.
3. Cover PIR lenses.
4. Enter the primary access code plus the sensor test code.
5. Trip each sensor.
6. Count the number of transmission beeps (see manufacturer's documentation for minimum requirements). At this point, the system will confirm which sensor has been tested and whether or not it has passed.
7. Press STATUS when you think all the sensors have been tested.
8. The system will let you know if you missed any sensors. If you have, test all untested sensors now.
9. Exit sensor test.

If a sensor fails the test the installer will

1. Use an RF sniffer to verify that the sensor is transmitting. Locate sensors within 100 feet of the panel whenever possible. Panel range varies with the installation environment. Mounting sensors within 100 feet of the panel reduces the impact of environmental conditions that may exist on the premises. Sometimes merely changing the sensor location can help overcome adverse premises conditions.
2. If necessary, improve sensor communication by repositioning the sensor, relocating the sensor, and/or replacing the sensor:
   - ☐ To reposition the sensor, rotate the sensor and test for improved communication at 90° and 180° from the original position.
   - ☐ If poor communication persists, relocate the sensor. To do so, test the sensor a few inches from the original position. Increase the distance from the original position and retest until an acceptable location is found. Mount the sensor in the new location.
   - ☐ To replace the sensor, test a working sensor at the same location. If the transmission beeps remain below the minimum level, avoid mounting a sensor at that location. If the replacement sensor works, contact the manufacturer for repair or replacement of the problem sensor.

## Testing phone communication

The installer will also perform a phone test to check the phone communication between the panel and the central station. A phone test takes a maximum of 15 minutes to complete. Before performing any phone test, the technician makes sure to set the communication lock.

To perform a phone test the installer follows these basic steps:

1. Enter the access code and the phone test code. The panel will indicate that the phone test is on.
2. Wait for a signal that the phone test is complete or that there has been a failure.

If there is a phone test failure:

1. Check to be sure the panel is plugged into the RJ-31X jack.
2. Reenter the access phone test codes.
3. If the phone test still fails, check to be sure the phone number that you programmed is correct. If necessary, change the phone number and enter the phone test command again.
4. If the phone test fails again, check the phone connection wiring.



## Testing communication with the central station

After performing the sensor and phone tests, the installer tests the system to verify that the central station correctly receives alarm information from your system. At this point, he or she will also verify that the X-10 lamp modules are working correctly.

To test communication with the central station the installer will:

1. Call the central station and tell the operator that you will be testing the system.
2. Arm the system.
3. Trip at least one sensor of each type—fire, intrusion, etc.—to verify that the appropriate alarms are working correctly.
4. If X-10 lamp modules are installed, check to see that they operate correctly. The lights should come on and stay on during the fire and auxiliary/medical alarms, and flash during intrusion alarms.
5. After testing the system, call the central station to verify that the alarms were received.

## Troubleshooting the system

For the installer, correcting common installation errors during a wireless installation is a matter of reviewing details that might have been overlooked during the installation process. Fortunately, the common errors are easy to correct. They are usually grounded in small oversights that require only a bit of adjustment.

The planning checklist below is an overview of common installation errors on the Commander 2000. However, these errors are typical for any brand of security equipment.

- ☐ The system won't arm.
- ☐ Central station is not receiving reports.
- ☐ Panel does not respond to hardwire input activation.
- ☐ Interior sirens are not producing sounds.
- ☐ Interior sirens produce low-volume alarm and high-volume status sounds.
- ☐ Panel does not power up.
- ☐ Power LED is flashing, the trouble light is flashing, and after pressing STATUS, the panel announces, "System battery failure."
- ☐ All panel LEDs are flashing.
- ☐ All panel LEDs are scrolling.
- ☐ Incoming voltage reading is 0.
- ☐ No dial tone on premises phone after wiring RJ-31X jack or connecting the DB-8 cord.
- ☐ Constant dial tone, preventing dial-out on premises phones.
- ☐ Phone does not work.
- ☐ Panel announces, "Sensor [*sensor* #] trouble."
- ☐ Panel announces, "Sensor [*sensor* #] failure."
- ☐ Panel announces, "Sensor [*sensor* #] low battery."
- ☐ Smoke sensor beeps once every minute.
- ☐ The panel does not respond to sensor activity. No alarm, chime, or sensor test sounds are present.
- ☐ The panel does not respond to touchpad commands.
- ☐ Lights controlled by X-10 lamp module do not work.

#### The system won't arm

1. If arming to Level 2, make sure all monitored perimeter doors and windows are closed.
2. If arming to Level 3, make sure all perimeter and interior sensors are closed.

#### Central station is not receiving reports

1. Check that the DB-8 cord is plugged into the RJ-31X jack.
2. Check for proper wiring of the RJ-31X jack.
3. Verify the phone number of the receiver line with the central station operator. Reprogram the phone number and retest if necessary.
4. Replace the RJ-31X jack.
5. Check that the DB-8 cord is properly wired to the panel terminals.
6. Replace the DB-8 cord.

#### Interior sirens are not producing sounds

Check for correct wiring at both the siren and panel terminals.

#### Interior sirens produce low-volume alarm and high-volume status sounds

Reverse the interior siren wires at panel Terminals 12 and 14.

#### Panel doesn't power up

1. Check the circuit breaker to be sure the circuit is live.
2. Check that the backup batteries are installed correctly, the battery bucket wires are connected to the panel, and the transformer is plugged in.
3. Check for proper wiring at the panel and the transformer.
4. Measure the incoming voltage at the panel terminals.

#### Power LED is flashing, the trouble light is flashing, and after you press STATUS, the panel announces, "System battery failure."

1. Check the circuit breaker to be sure the circuit is live.
2. Check that the backup batteries are installed correctly, the tabs are making contact, the battery bucket wires are connected to the panel, and the transformer is plugged in.
3. Check for proper wiring at the panel and the transformer.
4. Measure the incoming voltage at the panel terminals.

### All panel LEDs are flashing

This means the panel is in program mode. Turn off the mode.

### All panel LEDs are scrolling

A system alarm has occurred since the panel was last armed or the panel is in sensor test or phone test. Press STATUS for a description of the alarm.

### Incoming voltage reading is 0

1. Unplug the transformer.
2. Disconnect the wires from the transformer and the panel.
3. Check for continuity (i.e., a possible short) between any two wires or any open circuit on any wire.

### No dial tone for on-premises phone after wiring RJ-31X jack or connecting the DB-8 cord

1. Check the RJ-31X jack wiring.
2. Check the wiring from the panel terminals to the DB-8 cord.
3. Replace the RJ-31X jack.
4. Replace the DB-8 cord.
5. Perform a phone test after troubleshooting the phone line.



### Constant dial tone, preventing dial-out on premises phones

Polarity-sensitive phones exist on the premises. Reverse the wires you connected to the brown and gray wire terminals on the RJ-31X jack.

### Phone does not work

Disconnect the panel from the telephone jack. If the phone still doesn't work, the system is okay, and the trouble is with the phone or phone line itself.

### Panel announces, "Sensor [*sensor #] trouble."*

Replace the sensor's cover if it is off. Activate the sensor.

### Panel announces, "Sensor [*sensor #] failure."*

The sensor is not communicating with the panel.

### Panel announces, "Sensor [*sensor #] low battery."*

Replace the sensor's battery.

### Smoke sensor beeps once every minute

Batteries are low. Replace the smoke sensor batteries.

#### The panel does not respond to sensor activity/No alarm, chime, or sensor test sounds

1. Check that the sensor battery is installed.
2. Check the sensor battery for low voltage. Replace alkaline or lithium batteries if necessary.
3. Check that the sensor number is programmed into panel memory. Program the sensor if necessary.

#### The panel does not respond to touchpad commands

1. Operate touchpads from different locations within the premises to identify areas of intermittent operation.
2. Program the touchpads into the panel.

#### Lights controlled by X-10 lamp module do not work

1. Check that the lamp has a working bulb.
2. Confirm the lamp's operation at a working outlet.
3. Check that the lamps are plugged into X-10 lamp modules and that the lamp modules are plugged into outlets that are not controlled by a switch. Relocate to nonswitched outlets if necessary.
4. Check that the panel is powered by a four-wire line carrier power transformer and not a two-wire standard Class II power transformer.
5. Check that the HOUSE dial on the X-10 lamp module matches the house code programmed into the panel.

## In the next chapter

In this chapter you learned how a professional installs a security system. In the next chapter, we'll detail all the considerations that go into installing a DIY system, including how to perform a security survey of your home to determine what your system should include.

# Designing a DIY system 14

The goal of this chapter is to prepare you for installing your own system. Installing an alarm system yourself is certainly possible, but it should be approached with some caution. Unlike other DIY products, an alarm system is a life safety product that can call police, fire departments, and other emergency personnel. It is therefore very important that the system be installed properly to avoid false alarms, and worse, no alarms at all.

The tools required for most do-it-yourself security systems range from none to perhaps a screwdriver. Professionally installed systems tend to require a few more tools, such as an electric drill, wire cutters (for connecting the panel to a power source), a voltmeter, and a screwdriver.



Whether to buy a DIY system or to have a system professionally installed usually boils down to what you want to pay. Do-it-yourself wireless systems provide the same types of protection as professionally installed systems, but at a much lower cost. Kits ranging from $100 to $350 can detect intrusion through doors and windows and will sound an alarm to frighten would-be burglars away. In addition to kits, you can buy separate components for more complete door or window protection, along with glass-break security and motion detection.

Another cost advantage of buying a DIY system is that you can expand it at your own pace based on the amount of money you want to spend and when you want to spend it. Professionally installed systems can be expanded, too, but it might be hard to get an installer to come to your home for the addition of a sensor or two; the costs to the installing company might be too high to make the trip worthwhile. (Here is where having a small independent alarm dealer at your service is a big advantage. They compete with the national chains by giving excellent service and will make a call to your home just to put in an additional sensor.) If you're doing it yourself, you only have to pay for one component at a time if you like.

In DIY security system kits, most basic functions are preprogrammed at the factory, so access codes and delay times for entry and exit don't have to be set by the user. Programming has been simplified over the years even for trained technicians, yet it remains an obstacle for many consumers—much like the widely reported phenomenon of the blinking 12:00 on unprogrammed VCRs.

Nevertheless, DIY systems have a lot of appeal to people who don't want to spend more than a few hundred dollars on a security system. You can probably design a DIY system with the same number of components as a professionally installed system for about half the cost. That's not to say the components are of equal quality or that the system design will be up to professional standards; however, the amount of security will be approximately the same. (See figure 14-1.)

■ **14-1** *A do-it-yourself system touchpad.*



In addition, operating a DIY system is somewhat simpler than operating a professionally installed system. DIY manufacturers specifically design kits to be easy to install and easy to use because they know that's what consumers want. Keep in mind, however, that to make DIY systems easier to use, manufacturers include fewer features than would be found in professionally installed systems.

## Wired or wireless?

Once you've decided to go the DIY route, the next decision you'll need to make is whether to install a wired or wireless system. This decision can be a tough one, depending on the conditions under which you are installing the system. If you are building a home, for example, it seems obvious to install a wired system. However, this might not be a wise decision for everyone. Although the labor cost savings of a wireless system over a hardwire system are not as dramatic in new-construction situations, wireless systems can be easily expanded at a later time as your budget allows.

Just as with intrusion detection, the choice between a hardwire and wireless fire alarm system is the choice between paying for technology or paying for labor. Wireless components in a monitored system cost more than hardwire components, but the savings are always offset by the costs of installation. And when you look down the road to adding a new room or increasing your security protection, adding sensors to a wireless system is trouble-free when compared to the job of installing additional hardwire components.

If your walls are not open, installing a hardwire system should be avoided. We know excellent alarm companies who employ installation technicians with many years of installation experience, and even their technicians have trouble installing wires after the walls are up. Depending on conditions, it can take experienced technicians two to three hours to wire a single sensor. While carefully attempting to fish wires through walls and floors, it is very common for installers to cause damage to windows, roofs, electrical wires, plumbing, and floors.



One way around fishing wires through walls would be to staple wires down the outside of wall and around the baseboards of your home. However, this is an unsightly alternative. In addition, stapling wires can also cause false alarms. Staples always run the risk of piercing the thin jacket covering the wire. If the jacket is pierced, false alarms will eventually occur, and tracking down the wire and the precise location of the problem is very difficult.

Last, wires running throughout your home can act like an antenna. During thunderstorms, your system is in danger of damage from lightning. When you install literally a mile or more of wire around your home, all returning to the central processor, damage from lightning storms can occur. Many a security dealer has been called out to homes to replace hardwire control panels damaged during a lightning storm. In fact, after a large thunder clap, you can frequently hear alarm systems false-alarm if you live in an area where they are prevalent. While not totally immune from problems caused by thunder and lightning, wireless systems have a greater resistance to damage and false alarms because there is no network of wiring connecting the components together.

## Performing a security survey

When purchasing a DIY system, it's important to take the same steps you would if you were hiring a security dealer. You still want to do a complete security survey of your home, taking into consideration your family's needs and the amount of money you want to spend. A DIY package has a way of looking like a complete or adequate system, when in fact your situation might require a great deal more security than the basic kit contains. If you've only protected two doors and a window, and you leave six other windows unprotected, you're still vulnerable to undetected intrusion. Conducting a security survey to determine how many sensors to add to your system will give you a clearer idea of what it will take to secure your home and how much it will cost.

If you were having a professionally installed system designed for your home, a security consultant would meet with you and walk through the premises to gather information about you and your family's lifestyle and habits. You should conduct a similar walk-through before purchasing system components for your home.

If your primary security concern is the safety of individuals, that will determine placement of some of the system sensors. If your primary concern is the safekeeping of valuables—collections, furniture, electronics, jewelry—then your system will have a slightly different design. Will persons from outside the home be in and out of your home? If so, they must be instructed in the use of the system, and must be given access codes and perhaps even their own devices such as keychain touchpads or magnetic access keys.



### Evaluating your home

Walk through your home and make a record of all places of entry and exit, including doors, windows, and skylights. Using the security survey located in the Appendix of this book, keep track of the number of zones you want to protect. In security, a *zone* is an area of protection. A door/window sensor placed on an interior door is a zone. A door/window sensor placed on another door in the same room is another zone. Although both devices are in the same room, they are considered separate zones because they detect intrusion at different points.

A *perimeter* security device is any device placed to detect intrusion through an exterior door or window. *Interior* security is a device placed to detect movement after the intruder has gained entry to the home. Plan for a door/window sensor for each exterior door and as many windows as you can afford. Like a chain that is only as strong as its weakest link, perimeter protection is as effective as your least-protected door or window that a person can fit through.

Also remember that you want to move about freely in your home without setting off a false alarm. The alarm system you choose must be capable of having perimeter sensors armed while the interior motion sensors are deactivated. The motion detector will be grouped as an interior zone and will be turned off when your system is in the STAY or HOME mode (perimeter sensors on, interior sensors off because persons are moving about inside).

Your concern during this phase is primarily cost. The actual installation of the wireless sensors—even if there are dozens of them—won't be much of a concern, since they can be in place in a matter of moments with screws or double-sided tape. The number of exterior windows can add up very quickly, however, and you might decide that you cannot afford to have all of them covered. Base your decision whether to add a sensor on the access and exposure of the window. First-floor windows that are protected by bushes are going to be the most attractive points of entry. Second- or third-story windows that overlook the street are obviously least likely to be entered by a burglar. Take note of the type of doors or windows you want to protect, as the type of sensor you put on each one might vary according to its design or function.



## Front door

Like everyone else, burglars first try the easiest way to get the job done, and then they try the harder ways. Accordingly, as many as 90% of burglars break into a home through a door, and many are able to gain access through unlocked doors. Ironically, trying a door is one of the most inconspicuous ways to break into a house. If a neighbor sees someone walk into a house looking like he or she belongs there, suspicions will not be raised.

Your first line of defense should be to install door contact sensors at all ground-floor entry points. Also, you should design primary entry points as "time-delay" zones. By installing motion sensors in this manner, you can enter your residence when the system is in the ON position and have a programmable amount of time to reach the arming touchpad. Once at the touchpad, you enter your access code number to turn the system off. (You can also program other doors you use as primary entry points, such as a kitchen door and a garage door, as time-delay zones.)

The motion detector, usually a passive infrared sensor, must be programmed as a *follower device*, which means that if the front door is opened first, the motion sensor will allow you the time to enter and disarm the system before going into alarm. If, however, someone is detected by the infrared sensor and the front door wasn't opened first, the alarm will sound. Because the infrared sensor follows the front door, the front door is a *follower zone*.

Positioning the infrared sensor is crucial to the performance of the alarm system. If your front door is located in a typical foyer area, the infrared sensor should be installed so it is pointing away from the front door and in the direction of the rest of the house. If you have stairs to a second floor located in the immediate vicinity of the front door, the infrared sensor should be placed on the same wall as the front door, usually in the corner, and on the opposite side of the foyer that the stairs are on.

Motion sensors come standard with a lens that detects in an approximate 90° angle and approximately 40 feet out. By placing the sensor as described, you will detect anyone who breaks in by cutting a hole in the door or by breaking through glass on or by the door. In addition, you will detect anyone who broke in through the back of the house and who is now in the foyer on their way upstairs.



### Windows

After the front-door area, proceed through your home in a clockwise or counterclockwise fashion, asking yourself the same type questions you did at the front-door zone. For instance, when evaluating a window in a particular room, ask yourself the following:

1. As you look out the window, is there a clear, unobstructed view of the window to passersby and/or neighbors?
2. Is it likely anyone would see a burglar tampering with the window?
3. Are shrubs blocking or partially blocking the view of the window? (If so, consider pruning the bushes.)
4. Is the window on the front, street side of your home?
5. Is there sufficient light on the outside illuminating the area of the window?
6. Is there a privacy fence between the front of the house and this window, effectively blocking view of this window from the front of your home?

If the window is on the second- or third-floor, also ask yourself the following:

7. Is there a roof surface reachable from the ground that one could stand on to tamper with this window?
8. Are there branches of a tree close enough and strong enough to support someone's weight as if an intruder attempts to gain entry to this window?
9. Is there a downspout, trellis, or anything else attached to the house in the vicinity of this window that would assist a burglar?
10. Are there any ladders outside of the house not chained and locked that could provide access to upper windows?

Windows can be protected in a number of different ways depending on your lifestyle, the type of window, the degree of protection you want, and the amount of money you want to invest. Window contacts, like door contacts, are the first line of defense. If you install a window contact, it will detect a window opening. It is not designed to detect someone breaking the window, but in most cases, a simple contact is all you need.

You could also install a combination window contact/glass-break detector. Specially wired full-size window screens are also available to replace your existing half screens. Every 4 inches of the screen contains a barely noticeable security wire, as well as a built-in window contact and tamper switch. Thus an intruder would be detected if he tried to remove the screen to get to the window or attempted to gain entry by cutting the screen. A nice secondary benefit of screen protection is you can leave windows open and still be able to detect an intrusion.



## Outside

After completing your survey of your home from the inside, look at it completely from the outside. Do you see anything that makes your home more vulnerable? Is there anything that would assist a burglar in entering your home, such as a pile of bricks lying on the ground, a garbage can within reach of a window, or an air-conditioning compressor under an upper window?

During the survey you must try to think like a burglar. How would you get into your home if you had to? Ask your teenager, if you have one. Teenagers often have to get in their own home through unconventional means when they lose or forget the house keys.

## Dos and Don'ts of Alarm System Installation

**DON'T** install an infrared sensor pointing toward a window if the window is 40 feet or less from the sensor. Infrared sensors can false-alarm if sunlight hits the lens of the sensor at a certain angle. Car headlights can also false-alarm the sensor.

**DON'T** point an infrared sensor directly toward a rapid heat-producing device. For example, if you had a wall heater or fireplace directly in view of the infrared sensor, the sensor might detect the rapid change in temperature on the surface of the heater and activate an alarm.

**DO** use a pet-alley lens on the infrared sensor if you have pets that move freely about the house while the system is armed. These sensors do not see pets moving about close to the floor. But remember, if your pet climbs onto furniture or goes upstairs, the pet could still set the alarm off.

**DO** use an audio glass-break detector if you have large sections of glass in an area of your home. Essentially, an audio glass-break detector listens for the sound of breaking glass. Higher quality devices have dual-detection sensors that listen for two events before alarming: breaking glass and the shock wave that occurs when glass is broken.

**DO** place arming touchpads within easy reach of your entry points into the home. Also install a touchpad in close proximity to your bedroom. Good security requires that touchpads be conveniently located. If they are hard to access, you'll find yourself not using the system as much.



## How many sensors do you need?

After you have completed a security survey, you will also have an important piece of information that will help you decide which system to buy: the number of sensors a particular system can handle. Some DIY systems can take an unlimited number of sensors. Because these systems are not fully supervised—that is, sensors don't communicate their operating status at regular intervals to the control panel—they can take more sensors than systems in which extra operating power and circuitry are used to keep tabs on sensor operation.

Because of their size limits, DIY systems are generally best suited to applications such as small houses, apartments, mobile homes, and other small structures. Once you get into protecting dozens of doors or windows, you'll be better served by a professionally installed system that is fully supervised and monitored. Such systems also might have convenience features such as light control, automation, energy management, and telephone control—not all of which are yet available in an integrated DIY security system—that make life in a larger home simpler.

## To monitor or not to monitor

You're probably not going to hear a manufacturer of a DIY security system say that monitoring is an absolute must for the best home security. They're after a particular segment of the consumer market—one that's not necessarily interested in spending the $20 to $25 per month for a monitoring service. They instead stress—and appropriately so—the effectiveness of detection in scaring intruders away. A burglar who breaks open a door or window at 2:00 AM and activates a screaming siren isn't going to stick around in the hope that no one will respond to the alarm. But even if the threat of tripping an alarm is effective in deterring a burglar, that's not a reason to forgo monitoring.



The very best time—as far as you're concerned—for a burglary to take place or a fire to break out is when no one is home. That way, you don't run the risk of being confronted by a burglar who might panic and assault you (which occurs in about 10% of break-ins) or of being overcome by smoke and dying in a fire.

An unmonitored alarm will not scare the fire away, and as far as the fire department is concerned, the fire is not burning unless they know it's out there. Similarly, a home left alone to a burglar and a siren is still vulnerable to theft and vandalism because, to police, an unreported crime simply isn't happening. In this respect, people with unmonitored systems rely on the hope that noise will scare the intruder away and that the intruder won't call your bluff. Bluffing is fun and exciting when you're playing games, but it is dangerous when the stakes can be as high as life and death.

## Fire protection

Fire is certainly as devastating and deadly as burglary. When installing an alarm system, you should install fire detection as well.

As mentioned in Chapter 8, the two basic types of smoke sensors are ionization sensors and photoelectric sensors. *Ionization smoke sensors* detect particles of combustion. An alarm is activated when smoke interrupts the electrical current inside the sensor. *Photoelectric smoke sensors* work by generating a beam of light. If the beam of light is broken, an alarm sounds. Photoelectric sensors are especially effective in detecting "cool smoke," such as that caused when upholstery is smoldering.

Before installing fire protection, you should decide where the sensors should be placed. According to N.F.P.A. (The National Fire Protection Association), smoke detectors should be located on all floors of your home, and specifically in sleeping areas. This means if you have a two-story house, at minimum, a smoke detector should be installed on the ceiling at the top of the second-floor landing and on the ceiling of the first floor, usually by the stairs.

Smoke rises, and your stairway acts like an enormous flue that sucks smoke and air upward. Because your air-conditioning return air vent also pulls air, you should install the smoke detector on the ceiling closer to it. For maximum protection, the best-possible scenario would be to install a smoke detector on each level of the home and outside each bedroom.



Heat detectors are also recommended for good fire protection. Install heat detectors in areas where you can't install a smoke detector because some smoke naturally occurs there or because it is especially dusty, for instance in attics, basements, furnace rooms, garages, water heater closets, storage rooms, and kitchens. The high heat in an attic would false-alarm a smoke detector; however, a 190°F heat detector installed at the ceiling of your attic won't false alarm even on the hottest of summer days, but will detect a real fire quickly.

In air-conditioned rooms install either a 135°F fixed-temperature heat detector or a rate-of-rise heat detector. The fixed-temperature detector looks for temperatures to reach 135°F, while a rate-of-rise detector looks for a rapid change in temperature.

## The components of a DIY system

**Console** The console is the brains of the system. It listens for transmissions from sensors and activates the appropriate annunciators (sirens and horns) or lights. If the console is equipped with or attached to a digital dialer, it also activates the phone call to friends, relatives, or a central monitoring station.

Consoles can be wall-mounted or simply set on a tabletop (figure 14-2). Some consoles will have a visible antenna for receiving the signals from transmitters. The consoles have to be plugged into a power source and, if monitored, will be connected to a phone line.

Linear

■ **14-2** *Do-it-yourself system consoles are typically placed on tables or countertops.*

**Door/window sensors** Door/window sensors are the workhorses of the system (figure 14-3). These magnet and reed switch devices detect when a normally closed window or door is opened. The radio transmitter, which sends alarm and status signals to the console, will either be part of the reed switch assembly or a separate device attached to the reed switch with wire. Sensors in DIY systems tend to be larger than those in professionally installed systems.

**Motion detectors** Passive infrared (PIR) motion detectors sense body heat by measuring changes in the ambient temperature of a space (figure 14-4). When a person enters a room protected by a PIR, the person's body heat registers with the sensor and activates an alarm.

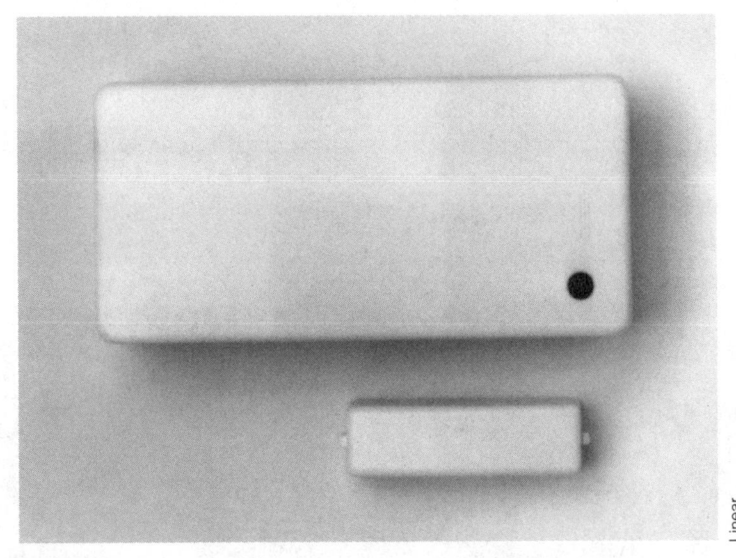

Linear

■ **14-3** *A wireless door/window sensor for a DIY system.*

Linear

■ **14-4** *A wireless motion sensor for a DIY system.*

They can be adjusted for sensitivity by setting them to alarm if any movement at all is detected, or by setting them to alarm only if two successive movements are detected within a certain time frame.

**Glass-break sensor** Depending on the technology, this sensor will either be designed to register the frequency of breaking glass or the presence of a shock wave from pounding.

**Panic button** A wireless panic button might be part of a touchpad or a keychain remote transmitter, or it might be a single button all by itself in a self-contained transmitter that can be clipped to a belt or worn around the neck (figure 14-5). A panic button can never be bypassed—it's always turned on—so in the event of an emergency, the system will go into alarm. When the system is monitored, the panic button may be set for police or for a medical emergency. In fully supervised systems, a sensor's identity and location can be determined from its signal. Portable panic buttons cannot be supervised, however, as their exact location can never be accurately indicated.

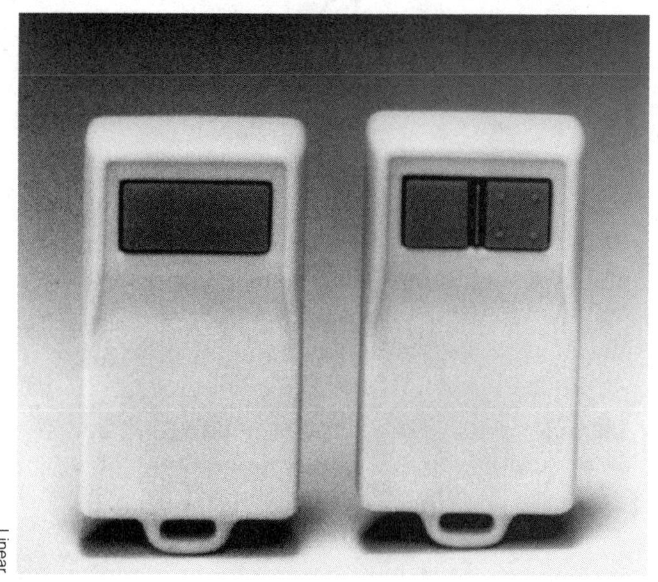

Linear

■ **14-5** *Remote controls for arming and disarming a DIY system. The two-button version also activates an emergency alarm.*

**Smoke sensor** Smoke and fire detection can be added to some DIY security systems (figure 14-6). In supervised systems, the precise location of a fire can be determined and the direction of its movement can be mapped as successive sensors go into alarm.

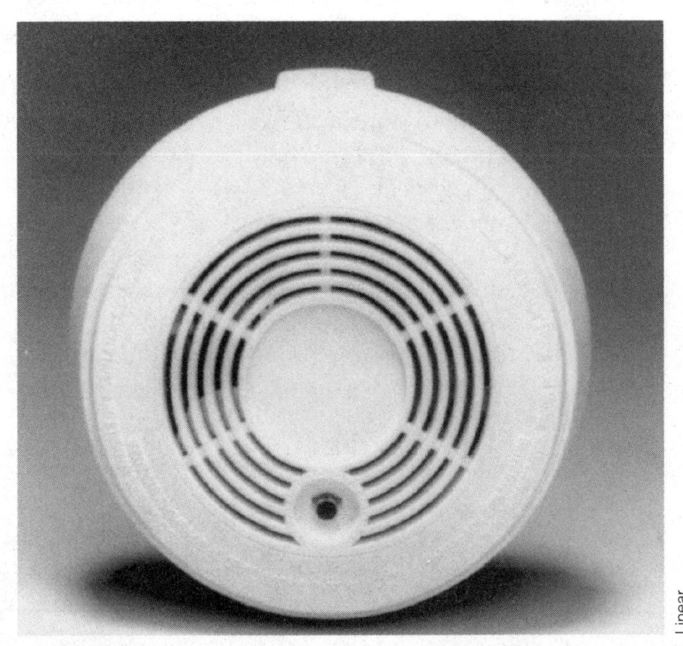

Linear

■ **14-6** *A wireless smoke sensor for a DIY system.*

**Sirens** In unmonitored systems, the siren is the hero. Once a burglar knows he's been discovered, he doesn't stay long—unless he thinks no one is home. But even then a burglar is at a psychological disadvantage—he's blown his cover—and will leave much sooner than when he goes undetected.

**Light control** Making a burglar think you're home when you are not is one of the most powerful deterrents to burglary. That's why many systems have light-control features that help you create that lived-in look when you're out.

**Carbon monoxide detectors** Many companies make stand-alone CO detectors that are either battery-powered or can plug into an outlet. Others make detectors that can be added to your security system. Research the cost of CO detectors and add a line to your system budget.

## Shoppers beware

Don't expect a lot of help from salespeople when you go shopping for a DIY security system. You can expect many of them to say, "I really don't know much about these things," and they'll be right. We know of one salesperson who tried to sell a customer a motion sensor without asking if she had a pet. The salesperson also did not foresee that the system would not allow the customer to disarm motion sensors at night if she wanted to move around the house without setting off an alarm. In another instance, the salesperson tried to sell door/window sensors and glass-break sensors without trying to sell a whole system, indicating ignorance of the basic functions of security system consoles and components.

Some displays will include a short video showing basic installation and how the system works. Despite their obvious intent to sell you a system, the videos are worth watching for what they show you about basic operation and installation.

You'll also get some idea of what the components will look like installed. Here we have a significant difference between DIY and professionally installed systems—aesthetics. DIY components tend to be larger than installed components. Part of the money you save when purchasing a DIY system comes out of the design budget, so you can wind up with a door/window sensor (the reed switch and magnet part) attached to the window and connected by wire to the radio transmitter, which is mounted on the wall. The more expensive professionally installed system will have the reed switch and transmitter integrated into a small component with no wires.



You can also expect to travel a few miles. DIY security systems are still not common enough for a store to carry a variety of systems. Instead, one chain might carry one system, and another chain carry one other system. The obvious shortcut is to call ahead and find out what each store carries.

Most important, do your homework first. Most likely, you will be dealing with a salesperson who doesn't know the difference between a motion sensor and door/window sensor or the difference between supervision and monitoring. Know in advance which brand you want and call only to ask if the brand you want is available.

## In the next chapter

Although DIY systems have the advantages of added convenience and lower cost, you might decide you want the additional security of a professionally installed system. Chapter 15 addresses how to select a competent dealer.

# 15 Choosing a security system dealer

Once you have come to the conclusion that you: a) need a security alarm system and b) want a professionally installed system, it is time to begin the search for a competent professional security alarm dealer. The relationship you have with a security company is different from your relationship with dealers of other consumer products. Your security dealer will walk through your home and learn a great deal about your security interests and your lifestyle. Accordingly, you're looking for a person and a company who you trust and who will be around for as long as you live in your home. They should be someone you can depend on for service, maintenance, and possible additions to your system as your home or family changes.



If your security dealer should go out of business, you can call on someone else for a system upgrade or for a service call. There are approximately 13,000 security dealers in the United States, and most metropolitan areas list dozens of companies. But buying security is a more personal matter than buying a refrigerator or stove. You are investing hundreds and sometimes thousands of dollars in a sophisticated electronic detection system that will link you with a service that dispatches police, fire, and medical authorities to your home within minutes of an alarm. You want a dealer who has a good relationship with alarm system manufacturers—that is, someone who can get the equipment he or she needs when they need it, someone who is knowledgeable about the technology available, and someone who is in business to install and service a system that will meet your particular needs.

Not all dealers carry the same alarm equipment. Some dealers specialize in wireless security systems; others specialize in hardwire. Some can install hybrid systems, which use a combination of wireless and hardwire components. Radionics, for example, manufactures a hardwire control panel that can be fitted with a wireless point expansion module that allows the addition of up to 17 wireless sensors.

Only authorized Radionics dealers can sell that particular system. Similarly, some alarm companies and manufacturers sponsor dealer programs that give authorized dealers the right to sell exclusive products. These programs trade exclusive territories, products, and services to qualified dealers as incentives to increase sales.

Another item to keep in mind when looking for an alarm dealer is that many independent dealers specialize in a certain type of security installation. Among those who specialize in residential security, some dealers might specialize in top-of-the-line narrow-band systems by ITI, Ademco, or Linear, or in spread-spectrum systems by Inovonics or Digital Security Products. Others might be experts in installing hybrid systems from Radionics or Napco. The rationale behind focusing on a particular brand of system is that it costs less to train installers on a particular system than on a variety of different systems. Installers are sent to costly manufacturer's training sessions. Many small companies can't afford to pay installers to attend several training courses or to carry large inventories of equipment from several different manufacturers. During the search process, ask if the dealer specializes in one technology or brand over another.



It is always a good idea to approach the purchase of an alarm system as an investment in an ongoing service. Be mindful of your comfort level with a dealer. Ask yourself if you would be comfortable having the company come back to your house for a service call or system upgrade. Look for cues to the dealer's sincere concern for your safety and security.

As mentioned, because no wires are needed between sensors and the control panel, it usually takes only one installer about a half day to install a wireless system (figure 15-1). Wiring will still be required to power the control panel, and a telephone link has to be set up between the control panel and your telephone line. Sensors are put in place with screws or adhesive tape.

Are you considering including home automation features? Many security dealers can also design home theater and audio systems, and some wireless security system control panels can control home lighting and heating/air conditioning. Even if you don't want to spend the money now on an automation system, you should choose a dealer who is authorized to sell the type of security system that has the control features you might want down the road.

Security alarm dealers run the gamut from small, start-up companies to 123-year-old worldwide organizations. In most any city, you

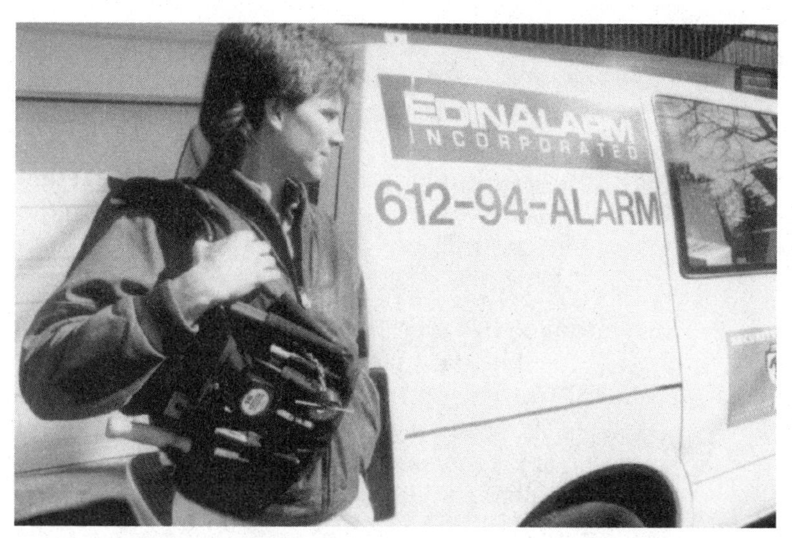


■ **15-1** *One advantage of a wireless installation is that one installer can usually do the job in less than a day.*

will have several reliable alarm companies to choose from. The challenge is to a) steer away from the dealers whose only goal is to separate you from your money; b) decide if you want to chance doing business with a relatively newcomer, given the failure rate of any new business; and c) decide on the qualities you would like to see in your security provider.



Thousands of companies are in the relatively new business of selling residential alarm systems. Since the first low-cost system was introduced by Brink's Home Security in 1983, many new companies have joined the ranks. The size of alarm companies is commonly reflected in the number of monitored accounts they have. A company like ADT has over 650,000 residential accounts, Westinghouse has 194,000, and Westec Security has over 65,000.

Small independent dealers, on the other hand, might have only hundreds or a few thousand accounts. But don't be fooled by the numbers. ADT might not be the right choice for you, even if they do have a thousand times more accounts than the independent dealer down the street. Large companies like ADT and Westinghouse are mass marketers of alarm systems. They can afford to offer wireless security systems at reduced rates.

You might have recently received a call from a telemarketer saying that there will be someone in your neighborhood demonstrating an alarm system that costs you nothing! You only have to sign a 36-month monitoring agreement. This is not a bad deal, since you have to pay a monitoring fee whether you buy a $1,000 system or a free one. At $25 per month for 36 months, the mass marketer is collecting $900 for a basic system of a control panel, two door/window sensors, one motion sensor, and 36 months of monitoring. Because they have sufficient capital, the mass marketers can pay their bills while waiting three years to turn a profit on the free system they installed. They make money by selling and installing systems in volume.

Keep in mind, however, that the mass-market companies make money by getting the system in as quickly as possible and providing as little customer service as possible. The faster the installer is in and out, the more money they make. So, from the mass marketer, you are getting basic security coverage, which is far better than having none.

In contrast, the local independent dealer is usually in the business of installing more complete systems and offering excellent customer service. Sure, a free system is cheaper than a $1,000 system, but the features and benefits of the two systems are very different. The cost of the more expensive system is created by the number and type of sensors and by the sophistication of the control panel's electronics. Higher-cost systems are often distinguished from mass-market systems and DIY systems by more sophisticated electronics such as built-in false-alarm protection, telephone control, voice feedback, temperature and light control capabilities, and many other features.



## Start-up companies

Start-up companies—those three years old or less—fall into two basic types. The most common are companies who have spun away from another company. Essentially, an installer or salesperson for one alarm company decides to go into business for him- or herself. But in some cases, such businesspersons are under-capitalized and overly optimistic about their ability to run a business. They might also mistake technical know-how for a good business sense. About 2 out of 5 alarm companies starting in this manner enjoy a certain amount of success. The other three, however, fail within 3 years.

The advantage of doing business with a new or small security alarm company is you might receive more personalized service than from a large national alarm company. The gamble is that the company will not have enough sales to survive in business. Although chances are excellent that you could find another company familiar with your system, the added hassle of having to do so is certainly not desirable.

The second type of start-up company is the entrepreneurial venturist—a company that is reasonably well funded, has expertise in starting and operating a business, and may or may not have expertise in the security alarm business. This type of company has a greater chance of succeeding because of its operating experience coupled with its ability to weather the tough times most new companies go through because it possess reasonable start-up capital.

You can tell a lot about an alarm company by the kind of salesperson they hire and what kinds of training they've had. It is therefore important that you check out the dealer carefully. It usually doesn't take long before a huckster shows its stripes. A call to the Better Business Bureau, the company's bank, and a couple of references will usually provide you with information upon which to base your decision.

## The experienced independent

Next comes the independent alarm company that has survived three to five years in business. The longer the company is in business, the less risk to you. Assuming the company has built a good base of recurring business, normal downturn cycles in business shouldn't be devastating.



Be aware, however, that some companies calculate their years in business in a way that isn't completely honest. For example, a company might advertise in the yellow pages that it has over 50 years experience. What the ad doesn't tell you is that the 50 years was reached by adding together the experience of everyone working for them, not by the actual years the company has been in business.

The differences between a 50-year-old organization and a 1-year-old organization are great. To protect yourself, check out the company thoroughly. You could, for example, go to your local library and check for the name in telephone books from prior years.

Ask for references, particularly of customers who purchased an alarm system from the company many years ago. Naturally, only a fool would give you the name of an unhappy customer. However, what you will find out will help you know what to expect from the sales and installation process.

You can also check with your local police and fire department. Ask what they know about the company. Don't expect them to recommend a company. Police and fire officials are usually not permitted to make recommendations. They can, however, tell you if they know of the company and if they know anything negative that might affect your decision.

## The authorized dealer

In between the independent alarm dealer and the regional or national dealer is the authorized dealer, who is a member of a national program (figure 15-2). Basically, this means the dealer is part of a national group of dealers, much like a State Farm agent in the insurance business. An authorized dealer program can deliver the best of both worlds. That is, the personal service and involvement of a locally owned and operated alarm company along with the support of a national company.

■ **15-2** *Companies who are part of a national dealer program will display their affiliation prominently.*

Like anything else, some authorized dealer programs are beneficial to the consumer, while others offer very little in the way of protection to the consumer. The difference is what the national company provides to you.

## National alarm companies

There are always advantages doing business with a large national organization. Their strong financial position usually means they're going to be around for awhile, which means you'll have someone to call if you have questions or need service. National companies also have buying power, which often translates into more competitive pricing.

Another benefit offered by many national companies is a free move of your system when you move from one residence to another. Companies like ADT Security Systems provide customers with a certificate that entitles them to a free move of the alarm system anywhere in the US. This benefit protects your original investment for years. (The free move is another advantage of wireless systems. Once a hardwire system is installed, it goes with the house—it's wired in. But wireless system components can be taken off the wall, packed up, and reinstalled in your new house. It's relatively inexpensive to offer the free move of a wireless system, whereas it would be impossible to promise to dismantle a hardwire system and reinstall it somewhere else.)

Unlike most independent alarm companies, national alarm companies almost always own and operate their own alarm monitoring stations, which are almost always Underwriters Laboratories approved. UL approval means that the monitoring station has a well-trained staff and that its computers, monitoring equipment, and the backup equipment meet rigid UL standards.

National alarm companies often operate multiple monitoring stations that keep duplicate records and provide redundant monitoring. If an emergency were to occur in your town that would hamper the ability of the local monitoring station to do its job, another one of the company's monitoring stations would take over. The net result is you are still protected.

An example of emergency backup monitoring happened in 1995 in Oklahoma City when the Federal Building was bombed. The ADT monitoring station located nearby was heavily damaged, as were other buildings in the vicinity of the Federal Building, yet no ADT customers lost monitoring. ADT simply switched all signals to another of their monitoring stations, and the monitoring of thousands of security alarm customers continued as if nothing happened.

## Comparison of services

If you were to break down the sale of an alarm system into its parts, it would look something like this:

- ☐ The design of the system by the sales alarm consultant.
- ☐ Installation of the alarm system.
- ☐ Ongoing service of the alarm system.
- ☐ Monitoring of the alarm signals.
- ☐ Ongoing consultative services.

Typically, large national alarm companies provide all of the above services through their own organizations rather than subcontract it. Independent alarm companies, on the other hand, provide some of the services, but not all. Independent alarm companies usually sell and design the system. Some install the system using their own employees, while others subcontract most installations. The same is true with ongoing service.

Monitoring, however, is another matter. Most independent alarm companies subcontract the monitoring of the alarm system to a wholesale monitoring provider for a portion of the monitoring fee you are charged.



## Working with your security consultant

Once you have a better feel for the type of security alarm dealer you want to contract with, the security consultant becomes the next criteria in your decision process. How professional, knowledgeable, and experienced he or she is makes an enormous difference in the final product.

The best equipment in the world installed by the best installation crew in the world won't work well if the system isn't designed properly, taking into consideration the type of home, the lifestyle of the people who live there, and the environmental challenges presented by pets. It is the security consultant's job to sort out these considerations and design a system that fits the family without making them a prisoner in their own home.

Professional security consultants take the time to ask lifestyle questions and determine your security needs. The consultant has to walk through your home to get the necessary information and to identify the areas most vulnerable to burglary. Once the walk-through is complete, the consultant will recommend equipment for the job of securing the home. The consultant should give a live demonstration to show you how the equipment works. A live demonstration will also tell you how user-friendly the products are.

In general, the easier the system is to use, the better (that is, unless the system is simple because it has an inadequate number of features). If the system requires complicated procedures or causes false alarms with the slightest mistake, you won't even want to turn the system on. Have the consultant show you how to turn the system on and off. Practice changing arming levels and ask how many different ways there are to sound an alarm. Ask to hear a siren go off so you'll know what an alarm sounds like.

A good system might still allow you to enter a duress code. A *duress code* lets the user signal an alarm while under the scrutiny of an intruder. For instance, if the intruder follows you into your home and demands you to turn off the system, you can enter a duress code that will deactivate the system, but will let the alarm company know you are in danger. But an experienced dealer will probably no longer encourage you to use a duress code because duress codes have been cited as one of the most common causes of false alarms. If your normal access code is 1234 and your duress code is 1235 or 2345, it's very easy to hit the wrong number or forget which number is which. Security information industry associations have recently made strong recommendations that dealer not offer a duress code—even if the system can be programmed for one.

While the consultant is there, learn how to bypass a door or window. The ability to bypass a zone comes in handy if you have a pool or like to spend time in your backyard. Bypassing your den door allows the remaining doors and windows to remain protected while still permitting access to your pool or yard through the den door. Think about it: when you are in a pool or hot tub, your ability to hear someone burglarizing your home through a side window or front door is severely hampered. As a result, you become more vulnerable. The correct use of a bypass feature lets you enjoy your home and yard while still having the benefits provided by an alarm system.

Keep in mind that it's the security consultant's job to learn as much as possible about how you live in your home, the dangers you're exposed to, and who will be using the system. It is also his or her job to conduct a thorough physical survey of your home with you or with you and your spouse in order to assess your needs. If the security consultant designs the system to fit his or her concept of what protection should include without your input, the consultant is short-cutting the process, and you'll wind up the loser.

**Consumer Tip: Five Tips toward Becoming a Demanding Customer**

In an era of niche marketing, security companies have to choose their customers and the best ways to satisfy them. Similarly, consumers have to choose their security dealers with great care. Selecting a dealer can lead to high blood pressure unless you're prepared to demand high standards for quality products and excellent service. Following are five tips to help you:

1. Be prepared for your first meeting with a security dealer by reviewing the key points in this book. Have some idea of the kinds of security you want. If you turn all the decisions over to the dealer—whose input should be extensive but not overbearing—your system will reflect the dealer's preferences rather than your own. This might not be all bad, if you find yourself without the time or interest in informing yourself about your security options. But you should understand that a decision to bow out of the process gives considerable control to the installing dealer.
2. Contact friends or neighbors who have gone through the process of buying a security system and hear what they have to say about the process—the high points and the low points.
3. Don't be afraid to say you have high expectations. You can reasonably expect a good security professional to spend a great deal of time listening to your security concerns, from the most urgent to the incidental. Get a feel for how demanding the dealer is on his or her own company, products, equipment, and employees.
4. Ask questions and expect answers. You might ask a question that the salesperson can't answer right away. There's nothing wrong with that. However, you can reasonably expect a willingness on the part of the company to get answers to all of your questions and get back to you on the ones they can't answer right away.
5. Listen for clear statements about what the dealers can do and cannot do. You can expect a reputable dealer to be candid. He or she should also be able to tell you how your own demands will affect the design and cost of the installation. Dealers who are not comfortable with a

particular demand should be able to explain why. There's no need for either party in the transaction to be defensive. Demanding people can be easier to deal with than more passive clients. By being clear about what you want, the dealer can be clear with you about costs.

## Licenses and insurance

Before committing to an alarm dealer, make sure the dealer has all the licenses required in your area. You can call city offices to determine what is required of an alarm company, or you can call the local chapter of the National Burglar and Fire Alarm Association (NBFAA) for the required licenses. When calling the local chapter of the NBFAA, ask them if they provide an alphabetical list of their local-area members. This could be an excellent way to locate reputable alarm dealers. Most, if not all, chapters of NBFAA set standards for membership in their organization. Most also police their own membership. The net result is a better, more responsible dealer.



Also check if the alarm company has all required insurance. In the alarm business it is standard for reputable alarm companies to possess a couple of types. First, most states require alarm companies to provide workmen's compensation insurance on all of its employees. However, because most states do not adequately enforce these requirements, it is highly possible for a small alarm installer to not have workmen's comp on all of its installing technicians. The problem is that if an uninsured installer working in your home is injured, you can become liable.

Accidents can and do happen in the process of installing alarm systems. This is especially true when installing wired alarm systems because the installer spends significant time in your attic fishing wires down walls. A missed step can result in the installer falling through your ceiling and hurting himself, and of course, causing damage to your property. If the company provides workmen's compensation insurance, the installer's work-related injuries would be covered.

Next is the issue of the damage caused to your home when the installer falls through the ceiling and lands on your antique dining room table. If the alarm company protects itself with adequate liability insurance, the damages will be paid for by the insurance company. If it is not insured and refuses to make good on the damages, your only recourse is to file a lawsuit against the company in hopes of recovering the damages.

While normal company liability insurance covers events that occur while installation is being done, errors and omissions insurance is required to protect you against future damages after the installer leaves your home. Post-installation damages are rare but can happen.

## The four-step method of buying a wireless system

With all of the above information in mind, we recommend a four-step method of buying a wireless security system. Taking these steps will help you make an informed decision toward a system and a dealer that is right for you.

### Step 1. Determine your security needs.

Take note of the security that is most important to you. How many doors and windows would you like to secure? Make a list of all points of entry to your home and assign a value according to their importance: *1* for doors and windows that must be secured (front door, back door, all first-floor windows, and other first-floor doors); *2* for second-story windows that are especially vulnerable to break-in (e.g., are hidden by trees or bushes, are near strong tree limbs, are in an especially dark area of the house, or provide outside entry to a room where valuables are kept, in use, or on display); and *3* for areas that probably do not need protection (windows overlooking areas where entry would be very conspicuous).



Try to determine a budget and your limits of flexibility. The size and scope of your security system will be based on the size and layout of the home, the cost of the components, and your personal feelings about security protection.

### Step 2. Understand your options.

Review the description of components in Chapters 7 and 8. There is a strong emotional side to the process of security protection, and decisions are best made under conditions that are least stressful. It's usually wiser to get the help of a security dealer as a preventative measure. If you wait until you have been burglarized, your fear might cause you to spend unwisely.

Consider whether you want to have local monitoring or if it doesn't matter to you where the central station is located. Which advanced features do you want? When do you want the system installed?

**Step 3. Choose three to five dealers.**

Once you know what equipment you want and how much you want to spend, search for a dealer who carries the equipment and has the staff to install the system in a timely manner. Be prepared to ask questions when you make your phone calls. Having direct, clear questions prepared in advance will let the dealer know you're knowledgeable about the options. Refer to figure 15-3 for a list of questions you can ask on the first phone call.

**Step 4. Meet with salespeople.**

Meeting with several different companies' sales representatives will tell you a lot about each company and what to expect during installation. It's also a chance to ask the questions that weren't fully answered over the phone. You'll see several differences among the salespeople. Does the salesperson come in, merely count the doors and windows, and give you a price? Or does he or she show a willingness to discover your particular security needs?



Don't be suspicious if a salesperson offers to let you try a wireless system free of charge on a trial basis. They can afford to give you a free trial because the system is so easy to set up. But don't allow anyone to place a yard sign at the front of your house until you have purchased a system. This is an old tactic to discourage competing alarm companies with whom you have appointments. They'll drive up, see the yard sign, and leave, thinking you have already decided on a system.

## In the next chapter

If you use the techniques in this chapter, you're certain to find a qualified dealer. However, as with any business, there are people who want to take advantage of you in the hopes of making a quick buck. Chapter 16 will address how to avoid bunko security scams.

■ **15-3** *Dealer Selection Checklist.*

- ☐ Is the company a national, regional, or local company? Is it part of a national dealer program?
- ☐ How long has the company been in business under their existing corporate name?
- ☐ Who in the area is a customer of the alarm company?
- ☐ Does the company specialize in residential security, commercial security, etc.?
- ☐ How long has the security consultant been in the business of security design?
- ☐ What does the Better Business Bureau say about the company?
- ☐ What do the local police and fire departments know of the company?
- ☐ Does the company possess errors and omissions liability insurance?
- ☐ Does the company possess workmen's compensation insurance?
- ☐ How do the references rate the company in the following areas:
  - ~ Professional installation.
  - ~ After-the-sale user training.
  - ~ Repair service timeliness.
  - ~ Monitoring services
  - ~ Overall customer service after the sale.
  - ~ Fulfillment of promises made at the point of sale.
- ☐ Does the company use subcontractors for installation and service or its own employees?
- ☐ Are they licensed to install alarm systems in this area?
- ☐ What kind of warranties do they provide?
- ☐ Are their installers certified by the National Burglar and Fire Alarm Association? (NBFAA)
- ☐ Is the company a member in good standing with any of the local area associations, such as:
  - ~ The local chapter of the NBFAA.
  - ~ The national chapter of the NBFAA.
  - ~ The local chamber of commerce.
- ☐ How long have they been in business?
- ☐ How many wireless systems has the company installed?
- ☐ How many monitored accounts do they have?
- ☐ Does the company have its own central station?
- ☐ If not, who will monitor my system and where is the monitoring station located?
- ☐ What kinds of training has the company's installers received?
- ☐ What kinds of wireless systems do they sell?
- ☐ How long has the company's installers worked for them?
- ☐ How many wireless systems has the installer installed?
- ☐ What are the names of three references?
- ☐ Do they have a referral program?
- ☐ Are they part of a dealer program that offers to reinstall a system for free if I have to move?
- ☐ What will the sales/installation process entail?
- ☐ How many installers will do the installation and what kinds of identification will they have?

*Note:* If you really want to get a feel for life with a system, you can ask for a free trial. Some companies will allow you to have a basic system for awhile so you can see if it really makes you feel any safer. This offer takes advantage of the wireless system's ease of installation.

# 16 Hucksters in the security industry

Unfortunately, hucksters exist in all businesses. They prey on the trusting nature most of us possess. Seniors are often the prime target for an unscrupulous security salesperson because they might be intimidated by electronic products. Hucksters take advantage of people's insecurity by pushing them to make a fast decision during the first sales call.

Sales trainers almost always teach first-call closing, which is the technique of getting someone to buy during the first sales call even if they have strong reservations about signing on the dotted line. First-call closing is actually the right course for a salesperson to take if he or she is to be successful in sales. However, techniques vary, and when you're dealing with a product that can make the difference between life and death, there's no excuse for playing games.



## Sales techniques

Because crime and fire are serious problems that can be disastrous for the victims, a salesperson with integrity should have a sense of urgency about selling a product that can prevent his or her customers from being victimized. Most security salespersons feel a responsibility to convince people to make a decision right away so they can start protecting their homes as soon as possible and help prevent the unpredictable from happening. Salespeople push hard because they know most people don't believe they're going to be the next fire or burglary victim.

The hucksters of the world also push for sales, but for entirely different reasons and with entirely different techniques. A huckster's concern for the buyer will be minimal at best, and it shows in his or her behavior and style. A huckster's knowledge of security products is also minimal. With the primary concern on making the sale, the huckster has little regard for the proper protection and the needs of the customer.

Hucksters tend to sell by fear. Some show graphic pictures of dead people at the scene of a fire. Some even show pictures of infants and children who have been fire victims. The pictures, no doubt, are actual pictures of what can happen, but using them as part of a sales tactic is unethical because it undermines rational thought processes. Several State Attorney Generals' offices around the country have declared that sales practice illegal. Hucksters also sell equipment that is grossly overpriced and often install it improperly.

We have to add that among honest security salespeople you will find an enormous variety of styles and approaches to selling alarms. A person who comes into your home and counts the number of doors and windows and then gives you a quote isn't necessarily a huckster, but he's not the most conscientious security salesperson on the planet either. As we discussed before, representatives of the large national alarm companies have to sell quickly in order to keep their costs in line. They thrive on volume, so they'll probably not do an exhaustive survey of your home.

Many dealers like to recommend a system that will give you the most complete coverage and then let you decide what not to protect, based on your budget and other priorities. There's nothing wrong with this technique at all. In fact, a company with the highest integrity should give you the benefit of their expertise in recommending the most comprehensive system possible. Because of their experience, an alarm company can point out the areas of your home that are the most vulnerable to burglary. But everyone has limits, and it should finally fall to homeowners to decide how much security they are comfortable with and how much they can afford to spend on security.



Huckster are impatient. They want you to decide immediately—often while some special deal is available—so you'll be compelled to sign right away. They might even complain that they went to a lot of trouble to pay you a visit and can't afford to come back or make a follow-up phone call. Remember, you don't owe any salesperson anything based on what it costs them to visit you or make a phone call. Visits and phone calls are part of the costs of doing business, and you make no obligation to them by letting them into your home for a sales presentation. Be suspicious of anyone who uses guilt to get you to buy anything. A respectful salesperson will understand your need to think about your purchase before making a commitment.

Legitimate sales companies are in the business for the long haul. They depend on customer satisfaction and goodwill to carry them into the future. Consequently, they are interested in selling to the right people. They want the persons affected by the alarm system to be present during the sales presentation so everyone understands the functions and the limits of the equipment, and so everyone understands the costs of the system and of monitoring. A legitimate alarm company wants to give you what you need to make an informed decision. They will therefore ask for an amount of time that allows for a complete presentation and follow-up questions.

A good salesperson wants the deciding party present so they can hear the facts about the products and services. If all the decision makers aren't present, the person relating the information tends to only remember how much the product cost. He or she often forgets the reasons why the cost is what it is. Sometimes with the elderly, concerned sons or daughters or other relatives should be present.

## Avoiding hucksters

Following are a few tips to help you avoid being taken by security scams. Although not necessarily common, hucksters are usually easy to spot because of their overbearing manner.

**Tip 1** Beware the salesperson who doesn't support the idea of friends or relatives being present during a sales presentation. You might want to make the decision with a spouse, for instance, and that's not a problem. You should get no objections from the salesperson.



Huckster management subscribes to the philosophy that "buyers are liars." If the prospect says they'll think about it, they are lying to you. Management establishes a policy that says a salesperson who doesn't close on the first visit will not receive a dime in commission, should the customer buy on a later date. So the salesperson has everything to lose if they leave your home without a sale. Of course, this type of policy creates a very high-pressure closing environment. Salespersons working under the policy of no pay if the sale isn't closed on the first visit will stay in your home until you bodily throw them out. They'll try every tactic and ploy to get you to buy. They'll attempt to make you feel guilty for wasting their time. They'll work on your fear, your principles, and anything else that will get them the sale. If they can't close the deal, they'll call the office so their bosses can attempt to close the deal over the phone.

**Tip 2** Call the Better Business Bureau in your town before seeing any company in your home. Chance are good that the BBB has information on the company if they are unscrupulous. There are plenty of good, legitimate security alarm companies in every area and you should have no trouble seeking them out.

**Tip 3** Know your rights. Federal legislation was enacted some years ago to protect the consumer from predatory salespeople and companies. A mandatory three-day right of rescission entitles all consumers to cancel a contract without penalty within three days of signing the contract. (Some states have extended this to seven days; check with your state attorney general's office.)

Not only must the vendor provide you with this right to rescind, that right must be clearly spelled out in the contract you sign, in sufficient-size boldface type. (In other words, it can't be hidden in the fine print.) If they fail to disclose this right in the contract, you may cancel the agreement at anytime now or in the future.

But beware: hucksters have found a way around this law. They convince the customer to sign a waiver of the right to rescind by stressing how dangerous it would be to live three more days without protection. Any alarm dealer who asks you to sign such a waiver is not to be trusted.

Another way hucksters skirt the rule is to arrive the next day to install the system. In the sales presentation they will tell you it could take as long as two or more weeks before they will be able to install the system. This makes you relax because you believe you have time to sort through your feelings and check out the company. The next day the installer arrives, explaining that an opening in the schedule came up. Companies do this because they know people feel obligated once the system is installed and are much less likely to cancel the purchase.



**Tip 4** Don't let a company representative make you feel guilty. If you're not ready for the installation, tell them "no." If they push, you have more reason for being suspicious. Tell them "no," then spend some time investigating the company. Remember the old saying, "If someone walks like a duck, looks like a duck, and quacks like a duck, they're probably a duck"? Well, if someone looks like a huckster, sounds like a huckster, and acts like a huckster, they're probably a huckster.

## In the next chapter

In Chapter 17, we'll take a close look at how you can avoid one of the most potentially troublesome aspects of owning a security system: false alarms. Now that false alarms are normally caused by user error rather than defective equipment, you can eliminate these nuisance alarms entirely by using common sense.

# Preventing false alarms 17

About 20 years ago, at a meeting of the Louisiana Burglar & Fire Alarm Association, the room full of local alarm dealers turned their attention to the podium as the chapter president announced the first order of business:

"The City Council of New Orleans has announced," he said, "that it plans to vote on an ordinance that will fine alarm system owners $100 for every false alarm called into the police department."

The president's announcement put everyone in the room into shock. The alarm system dealers gathered on that day had to wrestle with the notion that they were in the business of selling a product that could cost their customers $100 every time it was used improperly. Common sense told them that if such an ordinance were passed, they might as well start peddling the plague.

Now fines for false alarms are common, and there has been little backlash from consumers about the threat of fines from false alarms. Why? Because a great deal about alarm systems—how they're regulated, made, installed, serviced, and operated—has changed dramatically in the last 20 years. Alarm industry associations have developed guidelines for manufacturers, dealers, and central station operators that reduce the number of false alarms. Manufacturers have designed systems that required two or more buttons to be pushed simultaneously to avoid alarms caused by simply bumping into a button. They designed equipment and worked together with central station operators to create verification procedures that allow an operator to be in voice contact with the homeowner to verify an alarm before dispatching police. Dealers have gone to great lengths to educate consumers about preventing false alarms.

## Causes of false alarms

False alarms have four main causes:

1. User error (75%)
2. Security equipment malfunction (10%)
3. Environmental causes (weather) (5%)
4. Unknown (10%)

Far and away, the majority of false alarms are caused by customer/user error. About 25 percent of user errors occur because of authorized entry without using a code. Another 25 percent happen when the alarm user arms the system and then moves around inside the protected area. Employees who are responsible for closing the business at night and turning on the alarm system are often careless in doing so, and the more turnover a business has, the greater number of false alarms they create due to untrained users. Homeowners using their system on a daily basis forget access codes, forget arming levels, or fail to secure all doors and windows before arming the system. Wandering pets, helium-filled balloons, and drafts that move plants and curtains cause false alarms triggered by motion sensors.



Some false alarms are weather- rather than user-related. Severe thunderstorms, for instance, can cause some systems to false-alarm when wind, thunder, or lightning vibrate loose-fitting windows.

Monitoring stations screen out as many as 90% of the alarms actually received and only dispatch police on the 10% or less they cannot verify. Alarm companies argue that if it wasn't for them monitoring the alarms, police would have to deal with 10 times the calls they are handling.

By educating alarm users about the problem of false alarms, and by leveling false alarm fines against the worst offenders, statistics show significant decreases in the number of false alarms in many communities. Another way of combatting false alarms is through improving technology. Today's systems are much better than just three years ago, and as pressure continues from alarm industry groups such as NBFAA, SIA, and CSAA, along with the larger alarm companies such as ADT, systems in the future will reduce the number of equipment and environment false alarms that tax police response in larger cities throughout the United States.

## Steps consumers can take

The best news about false alarms is that there are so many things a system user can do to prevent them. We've identified the most important steps to greatly reduce your chances of causing a false alarm.

1. The most common error is not knowing how to turn your alarm system off, combined with not knowing your identification number to cancel the alarm before dispatch occurs.

   Accidentally causing your alarm to sound isn't a problem unless police are dispatched. If dispatch can be avoided, you've merely conducted a test of your alarm system, which is good.

   When your system is installed, make sure you are allowed to program your own four-digit code number into the system. That way, you can choose a number you already know, such as your ATM number, the year you were born, the year you were married, or a former address you will never forget. Make sure other family members can choose access code numbers easy for them to remember.

   Pick a personal ID code to identify yourself to the system should you accidentally set off an alarm. Make sure everyone else does the same. If your ID code is a number, it can be the same as your arming code, or it can be different.

   Make sure the company who monitors your alarm system has a policy of calling the residence before dispatching authorities. Some do not. Calling first gives you the opportunity to identify yourself and say that you do not want or need the police.

2. Having your system professionally monitored is a crucial step in false dispatch prevention. More than likely, you will accidentally set off your alarm system at least once. One morning, you or someone in your family will get up thinking about the day to come, and as you open the front door to get the newspaper, your alarm will sound.

   If your system is connected to an alarm monitoring center and you turn off your system within a few seconds, the signal sent to the monitoring center will indicate that you canceled the alarm, and no police response will occur. At worst, the monitoring center will call you and ask for your personal ID code word or number. Once giving it, the call will be canceled and no police will be dispatched. Under these circumstances, all you have done is to test your system.

If, however, your system is not connected to a monitoring station, the problem worsens. If you have an alarm that sounds an external siren, you are asking a great deal of your neighbors by expecting them to call the police for you. If they call the police every time they hear your siren, chances are the police will be dispatched for dozens of false alarms. Eventually, they will stop calling out of embarrassment. But if they don't call when hearing your alarm, what good is having the system? Professional monitoring solves the problem. They will call to verify alarms. It's their job. They understand that an occasional false alarm is to be expected.

With monitoring you can instruct your neighbors to not respond to your alarm because the alarm monitoring center will take care of it.

3. Program your system with a delay time that is sufficient for you to enter your home and disarm it at the nearest touchpad. There is no point in having to rush. When determining the delay time, consider normal interruptions that might slow your pace, such as setting down groceries, the phone ringing, or a child tugging at your clothing.



Also program your system for sufficient delay time to leave the premises after arming the system. Programming the system with enough delay time is easy; the hard part is not getting distracted by something after arming the system. To prevent this, some systems beep at decreasing intervals as the end of the delay time draws near. After arming the system with a 60-second delay, for example, the system will beep every 4 seconds until the last 10 seconds and beep every second until the end of the delay period.

Some systems also have exit fault protection. Essentially, *exit fault protection* comes into play when you take longer to leave your home than was programmed in your system. The normal exit delay gives you time to get out of the house before the system is armed. However, say you take too much time to get out of the house and sound an alarm, and you don't hear the alarm because you are busy getting into your car and starting it up. If you set off an alarm under these conditions, you will not be home to tell the monitoring station it is a false alarm. Exit fault detection would indicate you have taken too long to leave and cancels the alarm to the monitoring center.

4. Most systems today provide a duress feature designed to give you a way to call for help silently if you are ever forced to turn off your alarm system by someone who has taken you captive in your home.

   If the attacker makes you turn the system off as your enter the house, you can do so using a special code number that deactivates the alarm *and* sends a high-priority hostage message into the monitoring center.

   A discussion about eliminating this feature is currently underway by alarm industry specialists, because duress-code false alarms have become a serious problem. Users often confuse the duress code with their normal access code and send in duress alarms by mistake. Even if a user discovers their mistake, a duress alarm—because of the very situation it is designed to thwart—cannot be canceled. In addition, because the duress feature requires a silent signal to the monitoring center, you have no way of knowing you messed up—until the police show up.

   The problem has become troublesome in large part because system users differentiate the duress code from the access code by only one number or by transposing two numbers. People can't remember which is which and wind up making the wrong choice.

   The solution is to make the number you use for hostage/duress code very different from your regular arm and disarm code. Ask your dealer whether the duress code is available with your system, or if, indeed, the feature is even available anymore.

5. Most systems provide two or three touchpad emergency buttons: usually, one for police, one for fire, and another for medical or other emergency. Like your heat and smoke detectors, these are 24-hour devices and cannot be bypassed.

   One way to reduce the false-alarm problem is to make sure everyone in your home understands how the buttons operate. Some alarm systems have dual emergency buttons mounted side by side. Both must be pushed simultaneously before an alarm will occur. This is a very effective way to minimize false alarms. Pushing one button by accident can easily happen. Pushing two at exactly the same time is highly improbable.

   The buttons are designed to make panic alarms easy, so no access code is necessary to activate a panic alarm. Your responsibility, in return for the convenience of the panic alarm, is to train all members of the family in the proper use of the panic buttons.

6. Pets pose a challenge that must be considered when designing your security system. Systems installed without consideration for movement of pets create many nuisance false alarms. However, a small- to medium-size dog can be allowed to move freely around your home even after the alarm system is on—as long as your system is designed with your pet in mind.

   Interior passive infrared sensors can be equipped with a pet alley lens. This lens provides a safe area at the floor level while still detecting people walking at normal height. Bigger pets can be worked around in the same manner, but the higher you raise the protection to allow for a big dog, the more room you give a burglar. Inform your alarm installer of your pet at the time you purchase your system. Also tell your security consultant if you plan to purchase a pet in the future.

   Cats present a more difficult challenge for interior protection. Because cats climb, it isn't unusual for a cat to reach the same height as standing humans. A seasoned security consultant might be able to find a way for the cat and an infrared motion detector to coexist, but it's usually safer not to use a motion detector when cats are present.

   Cats also have been known to hide in the house when you think they are outside. Shortly after you turn your alarm on and leave the house, the cat comes out to play and triggers a false alarm. Worse yet, when the cat hears the alarm, it might run under a sofa to hide. When the noise stops some 5 to 15 minutes later, the cat comes back out, and the alarm sounds once again.

   Birds, if confined to their cages, present few challenges. However, make sure the infrared motion detector is not aiming right at the birdcage. If it is, a false alarm is sure to occur.

7. False alarms can be prevented by understanding fully what causes certain protection devices to alarm. For example, passive infrared sensors detect rapid heat change in the area they are watching. If the sensor is installed too closely to a heating vent, an alarm will occur whenever the heat comes on.

   Infrared motion sensors should not be installed near any heat-producing object, such as a stove or fireplace, or near a window, where sunlight could stream in, warming the area. A room full of glass, such as a solarium, should not be protected with an infrared sensor. Instead, an audio glass-break detector should be used. In this situation, be sure to consult with a professional security consultant. He or she will determine if the environmental conditions of your home require dual-technology audio glass-break detectors or single-technology detectors.

8. Consider your lifestyle. The way you live and move about in your home has everything to do with the number of false alarms you might cause. For instance, does anyone in your home habitually get up during the night and roam throughout the house? If so, infrared sensors turned on while you sleep are not a good idea. You can design the alarm system so that you can turn off the infrared sensors overnight (the STAY mode).
9. Buy high-quality equipment. And buy enough of it. You're not doing yourself or your family much of a favor if you install a system that is unreliable or causes false alarms. Cheap, unreliable equipment is a headache. Remember these sayings:

   "There is nothing made by anyone that someone else can't make cheaper."

   "The bitterness of poor quality lasts long after the sweetness of low price has been forgotten."



10. Have your system installed by the best company you can find. Poorly installed alarm systems false-alarm more than professionally installed systems. Alarm systems are constantly changing. As technology advances, it is important for alarm technicians to continue their education. The NBFAA has training courses technicians can take designed to update their knowledge.
11. Whether you install your own system or have it installed by professionals, install a quality wireless security alarm system. If, however, you choose a wired system, make sure the wires run are properly soldered and taped. Poorly soldered and exposed wiring cause false alarms. Poorly stapled wiring causes false alarms. Old wiring has an increased chance of causing false alarms.

    Make sure service/maintenance people working on your system in the future properly solder and tape the wires they work on.
12. Know how to cancel an alarm, and take the time to cancel an accidentally caused alarm. And if you can't cancel the alarm

in time, be considerate of the police and fire officials who have to race across town to get to your home when they receive your alarm. Make everyone in your home—and everyone who uses your home—takes responsibility for the proper use of the security alarm system.

## In the next chapter

Up to this point, you've learned everything you need to know about why you need a security system, what type best suits your needs, who should install the system, and how to operate it effectively by avoiding false alarms. In the last chapter, we'll look at your lifestyle so that your habits will work in tandem with a home security system. We've also included a fire safety quiz.

# Planning for life safety 18

Life safety planning starts with making your home a less attractive target for the burglar. Burglars need cover. They do not want to be caught. So start by looking at your home from the viewpoint of a burglar.

If you were the burglar driving down your street, would your home stand out as an easy target? Are there easy ways into your home not easily noticed by passersby? Do you have bushes hiding the view of a window or door? Are your windows locked? What about all doors? Remember that the way to make yourself and your family less a target for burglary is to remove anything that make the burglar's work easier. Remove or trim bushes that hide doors and windows. Now that you've made sure the locks in your home are sturdy, make sure you use them.



Do you have ladders or other objects lying around unchained that would assist a burglar in reaching an upper window? How about your neighbors: do they have ladders lying about? Can someone reach a window by standing on a table or on your air-conditioning compressor? Always chain ladders on the outside of your home, and make sure your neighbor does the same. Secure tables and lawn chairs so they cannot be used to access an upper window. Do not leave tools or garden tools outside where they can be used to assist a burglar.

Do you have a privacy fence? If someone got into your yard, would the fence provide the cover for the burglar while he breaks in? Consider changing fencing so someone could be seen in the side or backyard. Shadowbox fencing with larger spaces do a good job of keeping pets in while still providing some view of who is in the yard. Iron picket or wood picket fencing is better yet.

If your neighbor was having his home painted or some other work requiring someone to climb a tall ladder during the day, would your neighbor advise you and other neighbors beforehand? If you were having work done, would you normally advise all or most of your neighbors before the work starts? We have met victims of burglary whose home was completely emptied by what appeared to be a moving company. The neighbors saw the truck arrive and watched the men empty the house. They didn't call the police because they assumed the neighbor, whom they hardly knew, was moving. Another victim we met said burglars gained entry into his home by cutting a 3-foot-wide by 7-foot-high hole in the side of his house. When the neighbors were interviewed by the police, they all acknowledged hearing the chain saw running but assumed the neighbor was having work done on his home.

Meet your neighbors. Get everyone to agree to inform each other when workmen are scheduled to be working. Then if you see someone in a neighbor's yard, climbing a ladder, or in any way tampering with the home, call the police immediately. Do not attempt to handle the situation yourself. Do not approach the workmen and ask why they are there. If the neighbor didn't inform you of the work to be done, call the police.

The primary reason people don't get involved is that they fear the embarrassment of calling the police when they see something suspicious and then finding that the activity is legitimate. So instead they do nothing. If, on the other hand, neighbors agree to inform each other when they see something suspicious, they can know whether an activity is legitimate before calling the police. The old adage, it's better to be safe then sorry, was never more true. Take the chance of a little ribbing for being overly cautious rather than making headlines as the latest victim.



Does your home have adequate lighting? Are your lights on a timer set to come on automatically at dusk? Are the exterior sides and back of your home sufficiently lighted so as to make hiding harder? Install lights on the exterior with timers to come on at dusk. Leave them on all night. Yes, it will cost money to do so, but your security and life are worth the added expense. Motion activated lights that go on when someone steps within a certain radius are particularly effective.

Do you arrive home late at night, after dark? Are there areas in the front of your home where someone could hide so you wouldn't see them as you approached your front or entrance door? If you noticed a stranger standing on your property when returning home after dark, would you continue doing as normal, or would you pass your home and go to a neighbor's home? Risk embarrassment rather than endanger your life. If something isn't exactly right when you return home, don't go in. Go to a neighbor's home for assistance.

Are you less concerned about burglary during the day? Do you leave any doors or windows unlocked during the day? Do you ever leave the house unlocked to go next door, down the block, or to the store? Today, daytime burglaries happen as often as night. Burglaries are more likely to occur between the hours of 9 AM and 3 PM because neighborhoods are often empty during these times; children are in school, parents are at work, shopping, or involved in civic work. This is a perfect time for a burglar to work without being noticed.

If you don't have an alarm system, pin windows and sliding doors so they are hard to open from the outside. However, don't do this if you have an alarm system and the window or door is connected to the system. In this case you want the burglar to open the window or door in the manner your system is designed to detect. That way, the alarm sounds and scares the burglar away.

## Fire safety quiz

Nothing is more devastating than a fire. But to be truly prepared for a fire takes education and practice. Take this quiz to check your fire safety knowledge:

### True or False?



1. If you were trapped in a smoke- and heat-filled house on fire, the air nearest the floor is safest.
2. Fire travels from object to object and from room to room.
3. If I were asleep in my home and a fire started, the heat from the fire or the smoke would awaken me.
4. In case of a fire, all members of the family should get out the best way they can and then go to any safe spot outside the burning home.
5. If I am in my bedroom on the second floor, and a fire starts, I should immediately open my door and exit the house as quickly as possible.
6. If I can't exit by the door, I should plan to open my bedroom window and jump to safety.
7. If I was asleep and the fire alarm sounds, or someone gets my attention awakening me, I should get up fast and leave the house quickly.
8. I should sleep with my bedroom door open so I can get out quickly in the event of a fire, and so I can hear my children better.

9. If my clothing catches fire, I should run quickly to another family member or neighbor so they can help extinguish the fire.
10. If a fire starts, I should attempt to put the fire out first, then call the fire department if I can't.

The answer to all of the above questions is False.

Don't feel too bad if you answered True to many of the questions; most people get these questions wrong. In fact, that is part of the problem. We have received very little training in the event of a fire emergency. When we were in school, we were subjected to numerous fire drills to teach us to exit a burning building in an orderly controlled manner. But what training have we received for a home fire emergency? What training have your children received in the event of a home fire emergency?

Because we lack training, we are subject to misconceptions and erroneous information about what to do in the event of a fire emergency, and our lives and the lives of our loved ones depend entirely on what we do should a fire start in our homes.

Let's take the time to go over each of the questions again, and this time, we'll explain what should be done and why.



1. If you were trapped in a smoke- and heat-filled house on fire, the air nearest the floor is safest.

   False. The air nearest the floor is coolest; however, that air is deadly. While heat and smoke do rise, when plastics such as nylons, Dacron, and other chemically produced products burn, they give off gases that are heavier than air and therefore sink to the floor. These gases are extremely toxic.

   The safety zone is approximately 18 inches above the floor, below the descending heat and smoke and above the deadly heavier chemicals that sink to the floor.

2. Fire travels from object to object and from room to room.

   False again. Heat and smoke rise in a fire. The heat rising to the ceiling reaches temperatures in excess of 500°F. As this hot air travels across the ceiling completely, it begins to descend, spilling under doorjambs and moving from room to room. When the extremely hot air comes in contact with another flammable object, a new fire starts. This could happen three rooms away from the initial fire. Therefore, a possible scenario could be that newspapers lying on the floor 2 or 3 feet away from the original fire source don't ignite, while light, highly flammable draperies on a window three rooms do.

3. If I were asleep in my home and a fire started, the heat from the fire or the smoke would awaken me.

   Unfortunately, this too is false. Descending heat and smoke produce the opposite effect. A person is often rendered unconscious from the smoke and heat. If you breathe in the super-hot air, your lungs will burn and death will result. This is why fire chiefs throughout the country have pleaded with the American public to invest in smoke detectors, and are now imploring owners of smoke detectors to check and change the batteries in them regularly.

4. In case of a fire, all members of the family should get out the best way they can and then go to any safe spot outside the burning home.

   False. An important safety training rule to adopt in every home is the rule to establish a family rendezvous point—that is, a place where the family members agree to meet in the event of a fire. This point can simply be the neighbor's front lawn across the street.

   This rule is important. Say, for instance, all families members manage to get out of a burning home; however, since they scattered, one of them, perhaps a small child, is unaccounted for. The father or mother goes back into the home to search for the missing member and ends up dying in the fire. If you all meet at the same place, you can count heads and prevent the needless loss of life.

5. If I am in my bedroom on the second floor, and a fire starts, I should immediately open my door and exit the house as quickly as possible.

   False. If there is a fire in your home, opening a door from your bedroom to the hall or rest of the house can produce an explosion that could result in your death. (If you saw the movie *Backdraft*, you'll have a clear picture of this phenomenon.) The colder air from your room mixes with the hotter air on the other side of the door, and a backdraft explosion results.

   The proper way is to feel the door on your side first. Feel it high up. If it is hot to the touch, do not open it. If it doesn't feel especially hot, brace your foot against the bottom of the door and then open the door a few inches, just enough to feel the air on the other side. If you feel hot air, close the door and exit through your window. If not, you can proceed out of the door, staying low, so your head is about 18 inches off the floor.

6. If I can't exit by the door, I should plan to open my bedroom window and jump to safety.

   This is also false. Jumping out of window could result in death or severe injury. Instead, you should work your way over the sill and hang from the sill with your hands before dropping to the ground. If your second-floor windowsill is 12 feet above the ground and you are 5 feet, 2 inches tall, at full-arm extension hanging from the windowsill, your feet are only a little over 4 feet from the ground. Dropping 4 feet to the ground is safer than jumping 12 feet to the ground.

7. If I am asleep and the fire alarm sounds, or someone gets my attention by awakening me, I should get up fast and leave the house quickly.

   False. Getting up quickly is the problem. Because heat in excess of 500°F could be descending from your ceiling, sitting up in bed as you normally would to get out could very well put your head right into the descending heat. One breath could prove deadly.

   The correct way to respond to this emergency is to roll out of bed, gaining no height. Staying low, crawl to the door and feel it with your hand, as mentioned above.



8. I should sleep with my bedroom door open so I can get out quickly in the event of a fire, and so I can hear my children better.

   False. A closed bedroom door slows down a fire. It can give you precious minutes to save yourself and the rest of your family.

9. If my clothing catches fire, I should run quickly to another family member or neighbor so they can help extinguish the fire.

   False. Stop, drop, and roll is the rule to live by should your clothing ever catch fire. Running only fans the fire, making it far worse. Stop, drop, roll, and live!

10. If a fire starts, I should attempt to put the fire out first, then call the fire department if I can't.

    False. You should not attempt to put out the fire. Every second in a fire is critical. According to the National Fire Protection Association, you have less than two minutes to escape a home on fire alive.

    Calling the fire department before you get out of the house wastes valuable time and subjects you to possible injury or death. Get out first, then call the fire department.

# Glossary

**24-hour sensors** Any sensor in a system that cannot be disarmed or bypassed, such as fire detectors.

**alarm** A device for signaling an emergency.

**access code** A numeric combination that grants access to the system.

**AM** Abbreviation for *amplitude modulation.* This is where the message is impressed onto the carrier by changing the intensity of the carrier signal.

**antenna** A transmitting device that converts radio frequency energy (in a wire or cable) into a radio frequency electromagnetic field that travels through space, or a receiving device that converts an RF electromagnetic field in space into RF energy that can travel via a wire or cable.

**arm** To activate a security system so that its sensors detect changes of state and report those changes to a control panel.

**arming levels** A set of instructions that determines which sensors are activated and which are not. All sensors would be armed when you are away from home, for example, but interior sensors would not be armed when you are home and moving around the house.

**auxiliary/medical alarm** An alarm signaling the need for medical personnel.

**band** The frequencies used for a specific class of RF wireless communications system.

**bandwidth** The amount of radio spectrum used by a communications system. Both the transmitter and the receiver have a measurable bandwidth (measured in hertz). A receiver's bandwidth must be as large or larger than the transmitters bandwidth, so that all of the signal is received.

**baseband** The essential low frequencies of a message data-packet without the radio carrier.

**buddy system** An arrangement between two neighboring systems in which a cut phone line in one system activates an alarm via the neighboring system.

**bypass** An instruction to a security system not to arm a specific sensor.

**central monitoring station** An agency that receives alarms from subscribers' security systems and requests the dispatch of fire, police, or medical authorities.

**control panel** The "heart" of the security system. Receives transmissions from sensors and communicates data to the central monitoring station.

**crystal-controlled oscillator** An oscillator whose frequency is determined by a quartz crystal. These types of oscillators have the highest frequency accuracy and purity. The cost for these oscillators are higher than the other types, and their initial accuracy plus drift is better than 30 ppm (pulses per minute). These types of oscillators are used as the carrier frequency-determining component in security alarm transmitters of excellent quality, including narrow band and spread spectrum systems.

**decibel** A relative measure of the difference between the power level of two signals. The signal strength of a door/window sensor in a residential system might range from .1 mW for a close-in transmitter to .00000000001 mW for a transmitter in a remote closet.

**delay** The time allotted the system user to exit the premises after turning the system on without causing an alarm (exit delay); the time allotted the system user to enter the premises and turn off the system without causing a false alarm (entry delay).

**disarm** To turn the system or a sensor off.

**diversity** In order to eliminate the effects of multipath distortion, diversity communications systems use redundant signal paths between the transmitter and receiver. Two ways to implement diversity are called *spatial diversity* and *frequency diversity*. While odds of a deep fade occurring are generally very low, it occurs too frequently for the good reliability of a security system. With diversity, the redundant signal path is available during the occurrence of a fade on one of the paths. The odds of a deep fade occurring on two different signal paths at the same time is exceptionally remote.

**door/window sensor** A devise for detecting the opening of a door or window and for sending a signal to the control panel indicating the change of state.

**duress code** A numeric combination used to signal an alarm when forced to use the system by an intruder. An alarm goes out to the central monitoring station, but no local alarm is sounded.

**EPROM** Acronym for *erasable programmable read-only memory*. A chip that allows information to be erased under ultraviolet light and reused.

**exit fault protection** A feature on some alarm systems that prevents false alarms that commonly occur during exit delay times.

**false alarm** Any alarm sounded when there is no cause for alarm.

**fire alarm** A signal transmitted by heat or smoke detectors.

**FM** Abbreviation for *frequency modulation.* Creating a message that can be sent by a transmitter and understood by a receiver by changing the frequency of the carrier signal.

**free-air range** The line-of-sight unobstructed communications distance of a transmitter/receiver system.

**glass-break sensor** A device that detects frequencies that accompany the breaking of glass.

**handheld panic button** A transmitting device for activating a panic alarm.

**heat sensor** A device that detects heat or a rapid change in temperature that occurs in the presence of fire.

**home automation** The use of devices to automatically regulate or perform functions commonly associated with the home.

**interference** RF energy in the receiver's band that is not made by a system transmitter. Interference reduces the communications range of a transmitter-receiver system.

**interior sensors** Devices that register changes of state of interior spaces or doors.



**intrusion alarm** A signal transmitted by sensors that detect intrusion. Door/window sensors and motion sensors most commonly send intrusion alarm signals.

**ionization sensor** A sensor that detects particles of combustion. An alarm sounds if the electrical current conducted inside the sensor is broken.

**jamming** Caused when a signal from a transmitter that is not part of the system masks a desired signal from a system transmitter. Communications failure is the result.

**line seizure** The ability of security device such as a control panel to override other lines such as phone lines in the event of an emergency.

**magic key** A programmable disk that can be used for access control and for uploading and downloading control panel information.

**message data** The information sent by a communications system.

**modulation** The means of impressing the message data onto a carrier signal. Some of the common types of modulation are *amplitude modulation (AM), amplitude shift key (ASK), frequency modulation (FM),* and *frequency shift key (FSK).*

**module** In home automation, a device for interfacing controllers with security devices, lamps, appliances, and other devices. Each module has an address that may be unique or may be the same as other modules.

**monitoring service** See *central monitoring station.*

**motion detector** Also called *passive infrared,* or *PIR, detector.* An intrusion detector designed to register changes of temperature, which occur when an intruder enters a protected space.

**multipath fading** The damaging effects of multiple signals reaching the receiver at the same time. The result is reduced signal strength. If the signal strength is reduced below the receiver's detection threshold, the message will not be received. A good solution to the fading problem involves the use of redundant signal paths. Also called *nulls.*

**narrow band** Refers to a communications system that transmits and receives using only as much radio spectrum as is needed to pass the message data rate. In the security industry, *narrow band* has been used to mean any radio system that is not spread spectrum.

**no delay** With remote arming and disarming capabilities, the system user can approach the premises and, before entering, disarm the system with no delay. This feature eliminates the feeling of being rushed to turn off the system to prevent a false alarm after entering a delay door. A system that is armed with no delay will not give an intruder the benefit of the delay time to begin burglarizing the premises or assaulting the homeowner.

**oscillator** The circuit in a transmitter or receiver that converts the energy of the power supply into the high-frequency alternating currents often referred to as radio frequencies (RF). An oscillator is the circuit that produces the RF carrier.

**panic button** A remote control device or a button on a control panel that sends an alarm signal without requiring the use of an access code.

**panic pendant** A portable device that sends a wireless signal to the control panel. These devices can be worn around the neck with a neck strap, clipped to a belt, or placed in wall-mounted holders.

**photoelectric** A sensor that detects "cool smoke," such as that from a smoldering couch. An alarm sounds when the beam of light generated by the sensor is broken.

**radio carrier** A steady electromagnetic field having a known frequency and amplitude that propagates through space. In a useful communications system, the carrier propagates from the transmitter's antenna to the receiver's antenna without the use of wires. The carrier does not necessarily include any data or message but is capable of carrying data or a message.

**radio frequency** The alternating voltage and magnetic fields in a radio signal. Frequency is measured in cycles per second, which is usually referred to as a hertz after the man who discovered them. A megahertz (MHz) is a frequency of 1 million cycles per second.

**radio spectrum** The frequency range of electromagnetic radiation used for all RF wireless communications systems.

**receiver sensitivity** The ability of the receiver to hear the transmitter's signal. Sensitivity is typically measured at a signal strength so weak that the message is lost occasionally. This is the threshold of receiver detection and is measured in decibels relative to one milliwatt of signal power (dBm). A security control receiver with good sensitivity would have a detection threshold of –105 dBm. A receiver with excellent sensitivity would be able to decode signals at –112 dBm. The larger the negative number, the better the receiver's sensitivity is. Early garage-door opener and security alarm systems receivers using super-regenerative detectors would often have receiver sensitivities of only –80 dBm. This is about 1600 times less sensitive than a receiver having a sensitivity of –112 dBm.

**receiver** The RF analog and logic components in the security alarm panel that processes the signals received by the antenna and reconstructs the digital message of the transmitter.

**SAW resonator oscillator** An oscillator whose frequency-determining component is a surface acoustic wave (SAW) device. These types of oscillators have moderate to good frequency accuracy and moderate cost. Their initial accuracy plus drift is better than 1000 ppm (pulses per minute). These oscillators are often used as the carrier frequency-determining component in security alarm transmitters of good quality and recent design.

**sensor** A device that detects a change from one state to another. For example, a door/window sensor detects the change in a circuit from a normally closed state to an opened state.

**signal margin** The amount of excess signal available over and above the minimum required to receive a message. One way to establish the amount of signal margin is to divide the free-air range by the installed range. Some installers prefer this ratio to be greater than 10.

**silent alarm** An alarm received by a central station operator that does not activate a local alarm sound.

**smoke sensor** A device of either photoelectric or ionization design that detects the presence of smoke and sends a fire alarm signal to the control panel.

**soft wiring** Using existing wires in an application.

**spatial diversity** Using two or more antennas (on either the transmitter or the receiver) where the two antennas are switched. Having antennas with unique locations provides independent signal paths through space, allowing significant fade resistance. The antennas must be independent of each other and the receiver must use only one antenna at a time. If the antennas are only wired together, there is still effectively only one antenna, and therefore no spatial diversity or fade resistance.

**spread spectrum** A communications system where the transmitted signal is distributed in a specific manner over a bandwidth much larger than is required by the basic message bandwidth. Knowing the exact distribution pattern, the receiver gathers the RF energy and reverses the distribution process, thereby recreating a regular RF signal. The message is then recovered as in standard communications systems.

**supervision** The use of a special signal sent automatically from the transmitter to the receiver to inform the receiver that the transmitter is operating properly. If the receiver does not hear from the transmitter for a predetermined time, the control panel will flag the transmitter with a supervisory failure alert.

**transmitter power** The signal intensity of the transmitter. The higher the power radiated by the transmitter's antenna, the greater the reliability of the communications system. The electrical energy of the transmitter sent to the antenna is often rated in dB relative to one watt and is abbreviated dBW. The field strength of the signal radiated by the antenna is rated in volts per meter (V/m) or microvolts per meter (μV/m).



**transmitter** The circuit that includes logic that generates a baseband message describing the status of the inputs to the transmitter; a radio frequency oscillator that generates the carrier frequency, modulation method to impose the message onto the carrier frequency; and an antenna to radiate the signal.

**two-way voice** A technology for allowing a central station operator to hear what transpires after an alarm is activated and to talk to persons on the scene.

**watt** A unit of power, abbreviated as W. One thousandth of a watt is a milliwatt (mW). One millionth of a watt is a microwatt (μW).

**wavelength** The distance that the radio wave travels in one cycle of the transmitter's frequency. At 300 MHz, a wavelength is about 30 inches, and at 900 MHz, a wavelength is about 8 inches. An efficient antenna generally needs to be one-quarter or one-half the size of a signal's wavelength.

**wireless keychain touchpad** A system-controlling device designed to fit in a pocket or purse.

**zone** A sensor or group of sensors having a single numerical identity.

# Appendix

## Security Survey

Answering these questions will help you design a system that is tailored to your particular needs or interests. Use your answers to put the right system together, or show it to a security dealer and discuss the type of system you would like to have installed.

### Security Concerns

1. Have you ever experienced a burglary or fire?
2. Are you concerned about intrusion? At night? During the day? While away?
3. Are you concerned about personal safety? While inside? While outside?
4. Are you concerned about fire? At night? During the day? While you are away?
5. Do you have other security needs? Panic alarm? Furnace failure? Flooding? Power failure?
6. Based on your security concerns how much are you willing to spend on a system?
7. When would you like the system installed? What would be the best dates and times of day?

### Lifestyle Questions

1. How many people occupy the home or building?
   # of adults __________ # of children __________
2. Do you have pets or animals? Type? Size? Do they have free run of the house?

3. Is the home or building unoccupied very often? Vacations? Evenings? Weekends? Days?
4. Is any family member ever home alone? Nights? Weekends? After school?
5. Who has access to the home?

## Special Detection Needs

1. Which areas of the home are of most concern? Why?
2. What items are of special value to you?
3. Do you keep your valuables in one place?

## Security Audit

To determine the type of system that will suit your needs best, pencil in the locations of system components on this form.

### Security Control Equipment

Control panel Location: ______________________________

*Circle one:* *If yes:*

Wireless touchpads Y N Locations ______________________

______________________________

Hardwire touchpads Y N Locations ______________________

______________________________

Interior sirens Y N Locations ______________________

______________________________

Exterior sirens Y N Locations ______________________

______________________________

### Intrusion Detection Equipment

DS = Door Sensor WS = Window Sensor SS = Shock Sensor MS = Motion Sensor

PERIMETER DETECTION

| Item | Location | Item | Location |
| --- | --- | --- | --- |
| ______ | ______________ | ______ | ______________ |
| ______ | ______________ | ______ | ______________ |
| ______ | ______________ | ______ | ______________ |
| ______ | ______________ | ______ | ______________ |
| ______ | ______________ | ______ | ______________ |
| ______ | ______________ | ______ | ______________ |
| ______ | ______________ | ______ | ______________ |
| ______ | ______________ | ______ | ______________ |
| ______ | ______________ | ______ | ______________ |

INTERIOR DETECTION

| Item | Location | Item | Location |
|---|---|---|---|
| ______ | ______________ | ______ | ______________ |
| ______ | ______________ | ______ | ______________ |
| ______ | ______________ | ______ | ______________ |
| ______ | ______________ | ______ | ______________ |
| ______ | ______________ | ______ | ______________ |
| ______ | ______________ | ______ | ______________ |
| ______ | ______________ | ______ | ______________ |
| ______ | ______________ | ______ | ______________ |

## Fire Detection Equipment

*Circle one* *If yes:*

Smoke Sensors Y N Locations ______________________

______________________________

Heat Sensors Y N Locations ______________________

______________________________

## Misc. Detection Equipment

| *Circle one:* | | | *If yes:* |
|---|---|---|---|
| | | | Location |
| Portable panic buttons | Y | N | ______________________ |
| Freeze detectors | Y | N | ______________________ |
| CO detectors | Y | N | ______________________ |
| Others | Y | N | ______________________ |

# Cost Worksheet

List all system components here to estimate the cost of your entire system.

## Security Components

| | Number in system | Price each | Total |
|---|---|---|---|
| Control panel | | | |
| Wireless touchpads | | | |
| Hardwire touchpads | | | |
| Interior sirens | | | |
| Exterior sirens | | | |
| Smoke sensors | | | |
| Heat sensors | | | |
| Door/Window sensors | | | |
| Shock Sensors | | | |
| Glassbreak detectors | | | |
| Door/Window Sensors | | | |
| Motion Sensors | | | |
| Portable panic buttons | | | |
| Freeze sensors | | | |
| CO detectors | | | |
| Light modules | | | |
| Heat/AC control module | | | |
| Two-way voice module | | | |
| DIY package | | | |
| Labor/Installation | | | |
| Monthly monitoring | | | |
| Other equipment | | | |
| | | Total | |

## False Alarm Prevention Tip Sheet

Take these precautions to reduce the incidence of false alarms in your home.

1. Read your system instructions carefully.
2. Know the system well enough to teach someone else how to use it.
3. Know how to turn off the alarm after it has gone off.
4. Plan for all system users to be equally knowledgeable about the system's operation.
5. Watch the video user's guide if one comes with your system.
6. Check that doors and windows equipped with sensors fit securely in their frames.
7. Expect everyone with a key to your house to be knowledgeable about security system operation.
8. Know what to do in the event of a false alarm. Cancel the alarm as soon as possible. If the alarm has already gone out, contact your central station immediately.
9. Have your central station phone number handy at all times.
10. Have weekly alarm system reviews with all members of the family until everyone knows what to do in the event of a false alarm.
11. Have monthly alarm system reviews with all members of the family to see if everyone still understands how to operate the system.
12. Notify your central station if you are planning any remodeling or fumigating or other contract work that might cause interaction with your security system (sheetrocking, painting, wallpapering, phone line repair, floor sanding).
13. Do not point motion sensors at helium-filled balloons, drafty places where plants and curtains move, open windows, or places where headlights shine.
14. Before arming your system, lock all monitored doors and windows.

## Dealer/Salesperson Evaluation Form

Alarm company ______________ Name of dealer/salesperson __________

Date of visit/phone call ___________ Phone number ______________

Did the salesperson show proper identification at the door during his/her visit? ________________________________________

________________________________________

How many customers does the company have in the area? __________

________________________________________

What are the names and numbers of existing customers I can call?

________________________________________

________________________________________

________________________________________

Did the representative demonstrate how the system works? __________

________________________________________

Was the presentation clear? ________________________________________

________________________________________

Did the representative take time to listen to my expectations of the system? ________________________________________

________________________________________

Did the representative do a security survey of my home? __________

________________________________________

Was I encouraged to have all decisions makers present during the sales call? ________________________________________

________________________________________

Did I get a chance to see or try out the system components? __________

________________________________________

Have I come away with a positive feeling about the salesperson and the company represented? ________________________________________

________________________________________

Did the salesperson use any of the common huckster approaches? _____

________________________________________

Did the salesperson provide adequate and clear information about the terms of the sale? ______________________________

______________________________

Do I have a clear idea of the features and benefits of the proposed system? ______________________________

______________________________

Did the representative explain how long installation would take and is the installation time frame satisfactory? ______________________________

______________________________

Will the system be monitored in town or out of town? ______________________________

______________________________

Do I have enough information on which to base a decision I'll be happy with? ______________________________

______________________________

If not, what else do I need to know? ______________________________

______________________________

______________________________

______________________________

## Sensor Location/Programming Log

Use this log to register all the sensors in your system. You can also use this log to help program supervised sensors into your system.

| | | |
|---|---|---|
| Sensor # _____ | Type: __________ | Location: _______________ |
| Sensor # _____ | Type: __________ | Location: _______________ |
| Sensor # _____ | Type: __________ | Location: _______________ |
| Sensor # _____ | Type: __________ | Location: _______________ |
| Sensor # _____ | Type: __________ | Location: _______________ |
| Sensor # _____ | Type: __________ | Location: _______________ |
| Sensor # _____ | Type: __________ | Location: _______________ |
| Sensor # _____ | Type: __________ | Location: _______________ |
| Sensor # _____ | Type: __________ | Location: _______________ |
| Sensor # _____ | Type: __________ | Location: _______________ |
| Sensor # _____ | Type: __________ | Location: _______________ |
| Sensor # _____ | Type: __________ | Location: _______________ |
| Sensor # _____ | Type: __________ | Location: _______________ |
| Sensor # _____ | Type: __________ | Location: _______________ |
| Sensor # _____ | Type: __________ | Location: _______________ |
| Sensor # _____ | Type: __________ | Location: _______________ |
| Sensor # _____ | Type: __________ | Location: _______________ |
| Sensor # _____ | Type: __________ | Location: _______________ |
| Sensor # _____ | Type: __________ | Location: _______________ |
| Sensor # _____ | Type: __________ | Location: _______________ |
| Sensor # _____ | Type: __________ | Location: _______________ |
| Sensor # _____ | Type: __________ | Location: _______________ |
| Sensor # _____ | Type: __________ | Location: _______________ |
| Sensor # _____ | Type: __________ | Location: _______________ |
| Sensor # _____ | Type: __________ | Location: _______________ |
| Sensor # _____ | Type: __________ | Location: _______________ |
| Sensor # _____ | Type: __________ | Location: _______________ |
| Sensor # _____ | Type: __________ | Location: _______________ |
| Sensor # _____ | Type: __________ | Location: _______________ |
| Sensor # _____ | Type: __________ | Location: _______________ |
| Sensor # _____ | Type: __________ | Location: _______________ |
| Sensor # _____ | Type: __________ | Location: _______________ |
| Sensor # _____ | Type: __________ | Location: _______________ |
| Sensor # _____ | Type: __________ | Location: _______________ |

# Do-It-Yourself-Installation: Planning Form

1. Determine the purpose of the system.
   _____ Fire detection
   _____ Burglary detection
   _____ Environmental detection
   _____ Emergency response

2. Determine your budget for the project: $ __________

3. Plan the locations of hardwire components.

| *Circle one* | | *If yes, locations:* |
|---|---|---|
| Control panel | Y N | ____________________ |
| Hardwire touchpad | Y N | ____________________ |
| Hardwire sirens | Y N | ____________________ |
| Hardwire sensors | Y N | ____________________ |
| Two-way voice | Y N | ____________________ |

4. Plan the number and locations of wireless components.

| | | |
|---|---|---|
| Door/window sensors | __________ | ____________________ |
| Smoke sensors | __________ | ____________________ |
| Heat sensors | __________ | ____________________ |
| Sound sensors | __________ | ____________________ |
| Shock sensors | __________ | ____________________ |
| Glass-break detectors | __________ | ____________________ |
| Motion detectors | __________ | ____________________ |
| Environmental sensors | __________ | ____________________ |
| Touchpads | __________ | ____________________ |
| Panic buttons | __________ | ____________________ |
| X-10 modules | __________ | ____________________ |

5. Plan to explain the system to the user, including:
   - What the main system features are.
   - What the system limitations are.
   - How to operate the system.
   - How to bypass sensors.
   - How to turn the system on and off.
   - How to change system status.
   - How to test the system.
   - What the system sounds mean.

# Sensor Group Assignments

Sensors in some systems are assigned group numbers so that the control panel is able to identify the sensor's function in the system. Use this log to keep track of the sensor functions and groups in your system.

| Sensor # | Group # | Sensor Location |
|---|---|---|
| 01 | | |
| 02 | | |
| 03 | | |
| 04 | | |
| 05 | | |
| 06 | | |
| 07 | | |
| 08 | | |
| 09 | | |
| 10 | | |
| 11 | | |
| 12 | | |
| 13 | | |
| 14 | | |
| 15 | | |
| 16 | | |
| 17 | | |
| 18 | | |
| 19 | | |
| 20 | | |
| 21 | | |
| 22 | | |
| 23 | | |
| 24 | | |
| 25 | | |
| 26 | | |
| 27 | | |
| 28 | | |
| 29 | | |
| 30 | | |
| 31 | | |
| 32 | | |

# Index



Illustrations are indicated by **boldface.**

# E



# F



# G



# H



# I

## K



## L



## M



## N



## O



## P

## Q



## R



## S



## T

# About the Authors

Joseph Moses is Senior Editor at a security systems manufacturing company. He has published dozens of articles on wireless security system components, commercial and residential applications, system installation, and wireless technology. His articles have appeared in *Security Sales, Electronic House, Security Dealer, National Locksmith, Canadian Security*, and *Security Technology and Design* magazines.

Lou Sepulveda has managed several successful alarm companies in his 25 years in the alarm business. He conducts workshops and seminars on business management and acts as director of dealer development for a leading national alarm company. He has earned the designation of Certified Protection Professional, as granted by the American Society for Industrial Security. He is a regular speaker for the International Security Conference.